LOCAL POLICY FOR HOUSING DEVELOPMENT

Local Policy for Housing Development

European experiences

ROELOF VERHAGE
*OTB Research Institute for Housing, Urban and Mobility Studies,
Delft University of Technology
Faculty of Policy Sciences, University of Nijmegen*

Routledge
Taylor & Francis Group

LONDON AND NEW YORK

Contents

Preface   *vii*

1   Local policy for housing development and the residential   1
     environment
     1.1   Introduction   1
     1.2   Local policy and housing development   2
     1.3   The research project   9
     1.4   A micro-economic and an institutional perspective   11
     1.5   Influencing the residential environment   18
     1.6   Putting the pieces together   21

2   Cross-national comparison of housing development   31
     processes
     2.1   An approach to policy analysis   31
     2.2   A framework for a cross-national comparison   36
     2.3   About the research design   41
     2.4   Case study files: chronological overview   48

3   The incidence of costs and revenues   79
     3.1   The residual theory of land prices   79
     3.2   Case study files: financial analysis   88
     3.3   The cost side   108
     3.4   The income side   116
     3.5   Expenditure on the residential environment   120

4   Actors and activities, roles and relations   127
     4.1   An institutional analysis of the development   127
            process
     4.2   Case study files: actors, roles, and activities   132
     4.3   Unravelling the housing development process   148
     4.4   Power and dependence relations   157

4.5   Case study files: relations of power   162

4.6   Interdependence between the actors   179

5   Decisions about the residential environment   187

5.1   The analysis of decision making   187

5.2   Case study files: interactions and decisions   193

5.3   Realising the residential environment   215

5.4   The development process and the residential environment   233

6   Local policy and housing development: lessons to be learnt   237

6.1   Recapitulation   237

6.2   The value of the analytical framework   242

6.3   Changing the process to change the outcomes   248

6.4   Policy options for the Dutch situation   258

6.5   Where to go from here?   263

References   269

Documentation of the case studies   277

Informants   283

Appendix   285
Powers and practices for influencing housing development
in the Netherlands, the United Kingdom, Germany, and
France

1.   Introduction   285

2.   The Netherlands   286

3.   The United Kingdom   289

4.   Germany   293

5.   France   299

Preface

There are two reasons why the study of housing development intrigues me. We all live somewhere, and we all need some place for that. Questions of housing and land use are therefore of direct relevance to almost all of us. Let us call that the societal reason for my interest, as opposed to the second reason which is more academic. The development of housing can be described both as an economic process, and as a policy process. It is neither of the two, or both. To do right to this ambiguity, a framework for the analysis of housing development needs to combine different strands of theory, namely economic and institutional. That inevitably takes the researcher to the edges of the fields covered by these theories. It is these edges that offer the possibility to gather new insights. The subject of this study, the question of how we can influence the residential environment of new housing schemes, is only a small aspect of this field of study. But maybe a better understanding of this particular item might teach us some more about the intersection of housing and land use, and of policy and economics in a more general sense.

Carrying out the research that eventually resulted in this book was like making a long journey. The trips I made during the study, to some European countries are emblematic of this journey. These trips not only provided me with empirical data, but with much more. They provided me – through encounters with a lot of friendly people – with numerous perspectives, not only on how the data could be interpreted, but on many other things. An important result of my journey is the present book. It is up to the reader to judge whether it gives answers to the questions it sets out to investigate. But my journey had many more results than the book. Less tangible for the outside world, but not necessarily less interesting for me than the book you have before you now.

On my way, I met a lot of people that contributed – each in their own way – to what I found. First of all, there were the people who generously provided me with information about their land, or about the cases studied. Their names are mentioned at the end of the book. The research project that resulted in this book was funded by the Nederlandse Organisatie voor Wetenschappelijk Onderzoek (NWO, Dutch Organisation for Scientific Research). My new employer, OTB Research Institute, gave me room to finish my thesis and helped me with the maps and drawings. Martina Bernhard quickly but conscientiously read through the manuscript, tracing remaining errors (although errors that do remain are entirely my own responsibility). Some other people I want to mention in particular. Barrie Needham, Patsy Healey, Barry Wood, Claudio de Magalhães, Hartmut Dieterich, Benjamin Davy, Vincent Renard, Joseph Comby, Muriel Martinez, Frank Neidhardt, Willem Buunk, Tim Zwanikken, Marco Kerstens, Karel Martens, and the rest of a very fine group of colleagues at the Faculty of Policy Sciences at Nijmegen University: bedankt, danke, merci, thanks to all of you for your ideas, comments, and support. You made writing this book truly a European experience. And thank you, Claire, for... well, you know... everything. I invite you all to embark on this journey.

Roelof Verhage

"Un voyageur note ce qu'il trouve de singulier : s'il ne dit pas qu'il fait jour en plein midi à Modène, en conclurez-vous que le soleil ne se lève pas sur le quartier des jésuites ? Un voyageur note les différences ; entendez que tout ce dont il ne parle pas se fait comme en France.

Rien de plus faux que cette dernière ligne. Non, l'action la plus simple ne se fait pas à Rome comme à Paris ; mais cette différence à expliquer, c'est le comble de la difficulté. "

(Stendhal, Promenades dans Rome, *1827)*

1 Local policy for housing development and the residential environment

1.1 Introduction

In 1989 and 1991, the Dutch Ministry of Housing, Spatial Planning and the Environment published two reports that mark some important changes in housing development in the Netherlands. One of the two – *Volks-huisvesting in de jaren '90* (Social Housing in the 1990s) – initiated a more market-led housing policy, and a decentralisation of competences in the field of housing. The growing independence of the housing corporations since 1989 – and especially the grossing operation as of 1 January 1995 – was an important development that followed from this report. The other – *Vierde nota ruimtelijke ordening extra* (Fourth report on spatial planning extra) – indicated areas where housing development was to take place in the near future. As to the increasing importance of 'the market' in housing development, the proportion social sector/market sector housing in new housing development was changed from around 50:50 to 30:70. Since then, private developers that had been almost absent from the Dutch market for building land started to buy land and create strategic ground reserves, thus infringing upon the unique position the Dutch municipalities had had during the last decades in this market.

Whether the two reports caused this change, or whether there were other factors that brought it about goes beyond the scope of this book. Anyway, the Dutch municipalities were confronted with a new situation. Until then, they were used to buying the land required for housing development, then developing it and eventually selling serviced building plots. Thus the municipalities had an important and direct influence on the

1

residential environment: they were responsible for seeing that the development plan they had drawn up themselves was implemented. Moreover, because the municipalities sold the serviced plots, they received any possible value increase of the land due to the change from agricultural to housing use. Now, private developers had acquired the land before them, so these developers had become a new party to take into account. What would be the result of this change? How should the Dutch government react? Was there possibly a problem? One thing that was sure was that the municipalities lost a part of their influence. But would this be of influence on the housing development? Would another residential environment emerge? These questions lay at the basis of this study.

Although the occasion for these questions emerged in the Netherlands, their significance goes beyond the Dutch context. Insight into the relationship between local policy for housing development and the residential environment is increasingly important. Since the 1980s, the role of local government in urban development has changed (see for example Healey, 1992a; Albrechts, 1991). Following Goldsmith (1993: 66), the essence of this change can be described as the emergence of a local government whose main role is not to produce services, but to enable others to produce them. At the same time, cities in Europe are increasingly competing against each other to attract high quality activities. Good quality housing in good surroundings is an aspect of a city's attractiveness. Consequently, local authorities on the one hand are paying great interest to the residential environment, while on the other hand they are losing their direct influence on it. What will be the result of this and can local authorities continue to influence the residential environment? To answer these questions, more insight into the relationship between local policy for housing development and the residential environment is required.

1.2   Local policy and housing development

The local policy for housing development is pursued by the local planning authority. This is the (elected) governmental body that operates at a local level and that carries – as a public body – the responsibility for urban planning. In this chapter we deal in general terms with the role and position of the local planning authority in the field of housing development. To that aim the focus is first on the local policy for housing development: What form does it take? What is the role of the local planning authority? We argue that for the realisation of its policy aims, the local planning authority depends on other actors that play a role in housing development. This issue is addressed in order to make explicit the assumptions that underlie the subsequent analysis.

Local policy for housing development

The local policy for housing development generally covers such goals as the following:

- the need (or the demand) for housing should be satisfied at prices which (after taking into account possible subsidies) can be paid;
- the houses (and gardens) satisfy certain space standards;
- certain minimum local services (shops, primary schools, open space) are present;
- certain standards of urban design (to provide attractive living conditions) should be reached, as well as;
- a certain amount of social integration.

The local policy for housing development is a combination of several policies pursued by the local planning authority, i.e. housing policy, land use policy, and land policy. Housing cannot be built without land, so land supply is usually part of housing policy. However, the supply of land is affected by other types of public policy also. One is spatial planning, which affects how much land may be used for housing and where. The other is land policy, by which the public administration tries to realise its ideas about how land should be supplied, who should provide the necessary infrastructure, whether development gains should be taxed, etc. Under the common denominator of 'local policy for housing development', local planning authorities employ these policies to influence new housing development in four ways:

- by land use policy influencing the distribution and location of development;

3

- by the specification of standards for new housing development;
- by influencing land conversion and land development processes;
- by entering into working arrangements with private developers and housing corporations.

The local policy to influence the housing development is pursued by the local planning authority. But in urban planning and development, several actors play a role. Besides the local planning authority, these are for example private developers, landowners, and housing corporations. As with every other actor, the local planning authority has specific characteristics and a specific role. From its role as public body ensues the responsibility for urban planning. But in practice this does not mean that the planning authority can just impose its ideas on the other actors. Urban planning takes place in collaboration between all the actors involved in urban development. This perspective on the role of the local planning authority has emerged roughly since the start of the 1980s.

It resulted from an evolution in sociological thinking about 'collective action', of which the work of Crozier and Friedberg (1977) is an example. They describe how to understand the behaviour of actors in systems. The essence of their argument is that actors have a certain freedom of action in systems, but their actions are restricted and influenced by the system. As a result, 'collective action' cannot be understood using an approach where either the actor or the system seperately is taken as a starting point. There is a continuous exchange between the behaviour of the actors and the structure of the system. This is the central argument of Giddens (1984), who presents in his 'theory of structuration' a very elaborate analysis of the mutual relationship between what he calls structure and agency. That last observation is the starting point for the following analysis.

In such an analysis, the notion of 'power', as the inverse of dependence relations that exist between the actors, is central. Power exists only in interactions. As such, it is closely linked to the notion of negotiations. Actors in a system use their freedom of action to pursue their objectives. To do so, they have to enter into interactions with the other actors, because of the mutual dependence between them. In these interactions, negotiating powers are used to influence the outcome. The system influences the opportunities of the different actors by attributing

4

them more or less negotiating power. These sociological ideas have been introduced into policy analysis. For the way in which they shape the analysis of local policy for housing development, it is time to turn to the field of policy analysis, and more precisely to the field of urban planning as a specific form of public policy.

Local policy for urban development and the local planning authority

Traditionally, it is 'inherent in all ideas about planning and planning systems (...) that they must fulfil a regulatory role' (Ennis, 1997: 1938-1939). This can be explained as a logical result of the way in which urban planning emerged as a reaction to the uncontrolled urban growth during and following the industrial revolution. There was a great faith in the power of technology and the ability of human beings to transform their environment. Early planners formulated long term goals in the form of an end-state, which were to be reached by taking the steps defined by the planner. Ebenezer Howard's Garden City and le Corbusiers Radiant City are examples of these end-state – or 'blueprint' – plans. This view of urban planning influenced the actions of planners for most of the twentieth century. It corresponds to a view of the role of the planning authority as a central actor in a 'hierarchical policy field'. The local planning authority occupies this central position because it has responsibilities and powers that arise from its status as a public body. From its central position, the local planning authority 'regulates' the behaviour of the other actors.

From the 1980s onwards, this view of planning, and hence of the activities of the planning authority, has changed (see among others Albrechts, 1991; Healey et al., 1995; Lacaze, 1995; Ennis, 1997). It is not the subject of this book to describe why and how this shift took place. However, to understand the argument in the book, it needs to be clear what view of planning underlies the analysis. This is made explicit in the following, simplified way.

Hierarchic policy field.....................	........Interactive policy network
Regulation..	...........................Collaboration
Blueprint..	.....................................Contract

5

Imagine a continuum with the notions 'hierarchical policy field', 'regulation', and 'blueprint' on the left side, and the notions 'interactive policy network', 'collaboration', and 'contract' on the right side. The change of view about urban planning that has occurred since the beginning of the 1980s can then be seen as a gradual shift from left to right on this continuum. Of course, this is a very black and white representation of a discussion that is much more subtle (see for example Fischer and Forester, 1993; Teisman, 1992; Martinand and Landrieu, 1996; Healey, 1997). On the one hand, many more notions play a role, and could be placed on the continuum. On the other hand, placing these notions as the extremes on a continuum suggests contrasts that do not always exist and in any case are not so straightforward as this suggests. Notwithstanding these critical notes, the imaginary continuum with its three pairs of notions is used here to describe the approach towards the role of the local planning authority that is adopted in this book.

Hierarchic policy field/interactive policy network

There is a growing consensus in the field of policy analysis that a good way to describe decision making processes is by picturing them as an 'interactive policy network', as opposed to a hierarchic policy field. The idea of a hierarchic policy field corresponds to what we call above the traditional approach to planning: a top-down approach in which the planning authority is a central, steering actor that imposes its policy upon everybody, in the common interest. In an interactive policy network, the local planning authority operates as an actor in a network of interdependent actors that together realise the urban development. Because of the interdependence between the actors, no single actor is able to control the network. The activities and interactions between individuals, groups, and institutions influence the form and size of the network, which in turn influences the activities and interactions.

We take as a starting point that the local policy for housing consequences for our analysis; for the way in which the behaviour of the actors is interpreted (Dekker et al., 1992). It means that not the local planning authority and its policy, but the different participating actors and their objectives are taken as a starting point. Special attention is given to the distribution of power resources and the focus of the analysis is on the interactions between the participants.

Regulation/collaboration

In a hierarchic policy field, the way in which the planning authority influences the behaviour of others is by regulation. To put it negatively, it imposes constraints on the actions of others and thus influences their behaviour in the direction it has chosen. As a public body, it is entitled to do this. Although regulation is also used in an interactive policy network, it does not suffice to describe fully the way decisions are taken and influenced by the planning authority.

Regulation as a means of guiding decisions of other actors does not correspond with a situation in which the local planning authority and the other actors are mutually dependent, and where no single actor is able to control the network. Because of our approach to the planning authority as operating in an interactive policy network, it is necessary also to adopt another approach to decision making. In the following analysis, it is approached as a collaborative process (see Healey, 1997). This means that it emerges from interactions between all the 'stakeholders'. In housing development, the local planning authority is just one of these stakeholders. Others are private housebuilders, landowners, housing corporations. The local policy for housing development is influenced by all these actors, and in turn influences the behaviour and the decisions of all these actors.

This way of looking at decision making could also be explained using the notions of implementation and negotiation. The notion of implementation corresponds to a view of planning in which in a first phase a plan is made, which needs to be implemented in the next phase. That can work only if the maker of the plan – the local planning authority – is assumed to work in a hierarchic policy field. In an interactive policy network, the phases of plan making and plan implementation merge. The actors influence each other during plan making and plan implementation, hence the separation of these two phases becomes artificial. The process that takes place in such networks can be understood better as a negotiating process, in which the different actors mutually influence each other throughout the decision making about, and the realisation of, an urban development project.

Negotiations can be described as a means to secure the implementation of projects. This implies that a central actor – a planning authority – has set its objectives and now has to 'overcome the hurdle of implementation'. In this view, negotiations serve as a means used by the planning authority to overcome this hurdle. But in our view, policy making is not a matter of fixing objectives and then acting smartly enough to take all the hurdles to reach them. It is a continuous process of fixing, adjusting, and realising objectives.

Blueprint/contract

Traditional views of planning resulted in what is now called a blueprint planning, in which the planning authority defined an end-state of development in the form of a plan. The task of the planning authority was to choose the 'best' out of the different alternatives for development, i.e. the one that corresponded most to the public interest. 'The plan in effect embodied a comprehensive model of urban development strategy, providing instructions for public sector investment and guidelines for the private sector developer' (Healey et al., 1995: 3). This approach to planning puts the plan in the centre of decision making. All possible considerations that have been made concerning a certain spatial development are made in the process of plan making, and are crystallised in the plan.

In a situation where a number of mutually dependent actors work together, plans still play a role, but not as a blueprint. The publication of the plan is just one event in the long string of events that together compose the decision making process. Policy is made in the preparation of the plan, it is partly crystallised in the plan, but the plan is then used as a tool in further negotiations. Of course, it is a special tool. Plans are made by public authorities that have powers under public law. That means that they continue to play a role in the organisation of urban development, but in another way. When the urban planning process is seen as a negotiating process, plans are no longer blueprints. They are contracts in which the actors in the process fix the agreements they have reached, agreements that can be biased in the direction of one of the actor's interests. This can be explained by the power balance that existed in that particular decision making process.

1.3 The research project

We set out to analyse in which way local policy for housing development influences decisions about the residential environment. Above, we argued that the local planning authority should be seen as an actor that develops and pursues its policy for housing development in collaboration with the other actors in the housing development process. It does that in a decision making process that is greatly influenced by economic considerations, as the development of housing is also an economic process. On the one hand, the actors in the development process need each other, and they know that they need each other to bring a housing development process to a good end. On the other hand, the actors each have their own interests, and their freedom of action is limited by financial considerations. Thus we can distinguish two main considerations that influence the way in which the actors in the housing development process try to reach their objectives. These are the financial considerations and the power balances that emerge in the interactions between the actors.

However, to understand what happens when a housing scheme is developed, a view which concentrates only on these two considerations separately is too narrow. The question with which this book is concerned could very well be summarised in the words of Forester (1993: ix): 'What if social interaction were understood neither as a resource exchange (microeconomics) nor as an incessant strategizing (the war of all against all), but rather as a practical matter of making sense together in a politically complex world?'. The idea of actors making sense together in the politically complex world of housing development, in other words of actors actively coordinating their activities and interactions, is an idea thatis constantly present throughout the analysis.

Two central questions

The parties that play a role in the development of housing – local planning authority, private land developers and housebuilders, housing corporations, first landowners – have interests that do not always coincide. And yet, the parties depend upon each other. For example, private developers depend on the municipality for the granting of a planning permission. The

9

municipality often does not have enough money to realise a housing scheme, so it depends on the money of private developers or of landowners. This means that negotiations are going to take place, in which interdependent parties try to realise as many of their objectives as possible. Our assumption is that, for the outcomes of these negotiations, two things are of central importance. They form the central questions with which this book is concerned. The first one is:

1   *How much financial margin is available for investment in the residential environment and who receives it?*

As a result of housing development on a greenfield site, usually the value of the land on that site increases. This value increase could be used for investment in the residential environment. Seen in this way, local policy for housing development has two kinds of effects. On the one hand, it influences the amount of the value increase of the land. On the other hand it aims at reserving part of this value increase for expenditure on the residential environment. The first step in our analysis is to find out whether money comes available for the residential environment in housing development processes, and how much. This question requires a financial analysis of the development process, aimed at distinguishing the financial margin for expenditure on the residential environment.

But that is not all. Different actors play a role in housing development processes. According to the form that the process takes, one or several of these actors receive – in the first instance – the value increase of the land due to its development. The policy of the local planning authority to influence expenditure on the residential environment can therefore be expected to vary according to which actor receives this money. Moreover, each of the actors has different objectives as to what should be done with the money; so it can be expected that there are consequences for the expenditure on the residential environment, depending on who has the money. The part of a possible value increase that an actor receives is closely related to the activities in the process for which he is responsible. To find this out, an institutional analysis of the actors, their roles, and the interactions in the housing development process is carried out to answer the second central question:

2   How is the way in which the financial margin is used being influenced during the housing development process?

The local planning authority's policy for housing development can be seen as a way to influence the behaviour of other actors in the process. The local planning authority has a certain 'power' to implement this policy, and hence to influence the behaviour of other actors. But the other actors also have means to influence the behaviour of the local planning authority, for example by withholding their cooperation when they do not agree with the planning authority's policy. The analysis of the division of this power, and the way in which it is used is a central element of this study. Combined with the insights acquired by the investigation of the first question, this allows a better understanding of the way in which the residential environment is realised, and also how changes in the process of housing development might result in a different residential environment.

1.4   A micro-economic and an institutional perspective

To answer the questions above, we have to adopt a view of local policy for housing development that allows us to take account of the interdependence and the interactions between the different actors involved. Taking the 'housing development process' as the subject of study enables this focus on interdependence and interactions. For reasons that are explained in chapter two, we concentrate on greenfield housing development. This is the process during which a site on which there has not been any previous urban use is transformed from its original use to a housing use. Partly, this process is coordinated by market forces. But although housing development can be considered as taking place in a market environment, describing it merely as a market process would very much limit our understanding of it (see Lambooy, 1990; Van Der Krabben, 1995). Housing is supplied and demanded in a market that is only partly 'free'. To understand the functioning of the housing market, institutional arrangements must be taken into account. Therefore, we place the economic processes emphatically in an institutional context. Below, we explain what this means for our analysis.

11

The central question in any institutional analysis is how institutions affect the behaviour of individuals. However, this question can be approached in different ways. As Hall and Taylor (1996) observe, although the term 'new institutionalism' appears often in policy analysis, it is not always clear what is meant by it. The reason is that the same notion is used to cover different bodies of thought. Hall and Taylor distinguish between historical institutionalism, rational choice institutionalism, and sociological institutionalism. All approaches have in common that they address the relations between institutions and behaviour. The difference is in the way they do this. To describe the difference, Hall and Taylor distinguish between a 'calculus' and a 'cultural' approach to this relation. In a calculus approach, it is assumed that individuals behave strategically in order to maximise the attainment of their goals. In this approach, institutions '...affect behaviour primarily by providing actors with greater or lesser degrees of certainty about the present and future behaviour of other actor' (1996: 939). Without denying the purposive character of human behaviour, the cultural approach stresses that actors often do not act entirely strategically, but turn to routines to attain their purposes. In this approach, actors' behaviour is considered as 'satisficing', rather than 'maximising' utility. The role of institutions in this approach is to '...provide moral or cognitive templates for interpretation and action. (...) Not only do institutions provide strategically useful information, they also affect the very identities, self-images and preferences of the actors' (Hall and Taylor, 1996: 939).

The three types of institutionalism can be characterised by the different approach to the relation between individual behaviour and institutions they adopt. In the rational choice institutionalism, the calculus approach is worked out the furthest. The sociological institutionalism uses a cultural approach to explain the relation between institutions and actors. In historical institutionalism, a combination of both approaches is used, but this has led to 'less attention (...) to developing a sophisticated understanding of exactly how institutions affect behaviour' (Hall and Taylor, 1996: 950). Without going into detail, it can be concluded from Hall and Taylor's analysis that all approaches have their strengths and weaknesses. This led Ball in his review of institutions in British property research (1998: 1515) to the conclusion that 'If "the proof of the pudding is the eating", far more research work on property institutions is probably needed before firm conclusions can be reached on which institutional approach is best when studying specific aspects of property development'. This study does not attempt to find out which approach is 'best'. The inspiration for the empirical work is better described by Hall and Taylor (1996: 955) when they suggest that '... the time has come for a greater interchange among them'. In the research project that underlies this book, empirical data of a specific property development process – the development process of greenfield housing – is gathered and analysed. Hence as an offspin, this book aims at contributing to the knowledge about institutional approaches to policy processes.

The approach towards the development process used for structuring the empirical analysis comes closest to what Ball calls the ASH model (for Agency–Structure Healey), referring to a paper by Healey (1992b). In the terms of Hall and Taylor, Healey classifies this approach to the development process under sociological institutionalism. She develops '...an approach to the description of the development process which recognises the variety of agencies, agency relations, activities and events involved in development projects' (1992b: 33). To this aim, she distinguishes four levels through which the analysis of the development process should proceed. First a mapping exercise to describe the actors, agencies and events in the process. This forms the basis for the distinction of roles and power relations that evolve between them. On another level, an analysis of strategies and interests of the actors should highlight the driving forces between the behaviour of the actors. This could then be related to the resources, rules and ideas governing the development process. The fourth level of analysis links the process to the prevailing mode of production, mode of regulation, and ideology of the society in which the development is being undertaken.

This approach to the development process bears a close resemblance to the 'method of institutional analysis', developed by Ostrom (1986). She proposes that behaviour in 'institutional arrangements' be analysed by using the concept 'action arena' (Ostrom, 1986). An action arena includes a model of an action situation and a model of the actors in that situation. In

13

this research, a combination of the analytical tools provided by Healey and Ostrom is used to describe the greenfield housing development process. As to the driving forces behind the dynamics in the development process, ideas derived from sociological institutionalism are combined with insights from rational choice institutionalism. Thus, the suggestion of Hall and Taylor is taken up that the way to carry further the different institutional approaches is to favour the interchange between them.

The housing development process from an institutional perspective

In an institutional analysis, the 'institutional context' plays a key role. This context includes not only public sector interventions in the market, but also the composition of the group of actors, the different strategies of market parties, the institutional relations between market parties and the public sector, the impact of various 'rules' – not only legislation, but also norms and values – on market processes, etc. With a somewhat different wording, Burie described the housing development process in the Netherlands in this way as early as 1972 (1972: 40-49). According to Burie, the main actors or participants in the process are local government, brokers, designers, construction companies, private house builders, and housing corporations. The main roles the actors fulfil are those of administrator, initiator, designer, builder, and accommodating agency (that is responsible for finding occupants for the dwellings in the social sector). The combination of roles he fulfils characterises a participant in the process. Furthermore, Burie distinguishes actors that participate more indirectly in the housing development process, such as research workers, consultants, and finance companies.

Healey (1992b) describes the development process as a production process with as inputs ('factors of production') land, labour and capital and as outputs (besides profits, jobs and wider impacts) material values, bundles of property rights and symbolic/aesthetic values. Each development project involves accomplishing a series of events through which a site or property is transformed from one use to another. These events include identification of development opportunities, land assembly, project development, site clearance, acquisition of finance, organisation of construction, organisation of infrastructure and marketing and managing

14

the product. The events may vary in the order in which they are undertaken and who undertakes them. Together, the activities constitute the development process (Healey, 1992b: 39).

The role of the local planning authority in the development process is important for the outcomes of the process in two ways. The local planning authority is a participant in the development process in that it can fulfil roles that could equally be fulfilled by market parties. The local planning authority also sets – either deliberate or unintended – preconditions within which the housing development process has to take place (Van Der Krabben, 1995: 89). Local policy for housing development sets the framework within which the housing development takes place and thus influences the process of which the residential environment is an outcome.

A micro-economic approach

Above, the emphasis was on the way in which 'policy aspects' of the housing development process can be analysed. We now turn to more 'economic aspects' of this process. The policy for housing development is pursued in a market environment. However, in countries where land and housing markets are greatly influenced and regulated by public policy, it is clear that the development of new housing is more than a straightforward translation of demand into supply. It is a complex process in which different actors pursue different objectives. The (local) planning authority usually does not want the interaction between these actors to be coordinated only by market forces, for several reasons. Korthals Altes (1998) mentions the specific character of the land market (this is dealt with extensively in chapter three), the mutual influence of parcels on each other, and the public space which is always part of housing schemes. Nevertheless, the actions of the actors involved in housing development can be considered to work through the forces of supply and demand. This applies also to the effect of other institutional arrangements, or of social or cultural considerations. The price for houses can be described as resulting from the interaction between supply and demand, when both supply and demand are influenced by these factors (see for an example of such an approach applied to land prices Needham, 1992).

15

It is not the intention of this book to give an economic explanation for the functioning of the housing development process, but considerations regarding the interplay between supply and demand need to be addressed to analyse the behaviour of the actors in the process and its influence on the outcomes. Therefore, a financial analysis of the housing development process is also part of this study. This part of our analysis is constructed around the notion of the 'financial margin'. There are two sides to that notion. On the one side, there is the rational nexus argument. When housing is developed on a greenfield site, this usually goes together with value increases. The process of transformation of the land from agricultural to housing use generates an increase in value of the site. There is a strong argument in favour of using this value increase to finance the residential environment, i.e. the rational nexus argument. This is the argument that necessary expenditure which is caused by the housing development should be paid by the development. Expenditure on the residential environment is mainly provoked by the realisation of a housing scheme, and it mainly benefits the people in that housing scheme. Therefore it seems reasonable that the value increase that occurs when a housing scheme is developed is used for such expenditure to avoid charging the general taxpayer for such expenditure. The size of the value increase thus determines, together with possible subsidies, the size of a financial margin for expenditure on the residential environment in a housing development process.

On the other side, there is the observation that expenditure on the residential environment is restricted by what is financially possible: the financial margin in a housing development project sets the limits for expenditure on the residential environment. To spend money, you have to have income. In the development process this usually comes from selling the building plots and the houses. In some development processes it partly comes from subsidies from either local or higher authorities. Costs are made for – among others – the provision of primary and secondary services, normal profits for the developer, connection to infrastructure networks, contribution towards other residential development projects. Some of these items influence the residential environment. The money for providing them has to come from somewhere. The income and the expenditure in a development process have to be in balance. The way in

16

which costs and income are balanced – i.e. where the income comes from and on what items on the cost side it is spent – is specific for each process.

When described in this way, the notion of the financial margin appears to be quite straightforward. This must be qualified. When housing development processes are studied, a financial margin hardly ever appears. Costs of land acquisition, servicing of the area, and the realisation of facilities on the one hand, and income from the sale of building land and sometimes out of subsidies on the other hand, are always part of a development process and can be analysed. But these calculations do not show a financial margin in the way it is described above. If income is substantially higher than costs, this is integrated in the balance sheet. Activities are undertaken either to raise the costs, or to lower the income. This is dealt with in detail in chapter three. The point that is made here is that the notion of the financial margin is used as an analytical tool, to facilitate our understanding of how during the housing development process decisions about the residential environment are taken. In our investigation it is not seen as an item on the balance sheet of a particular project. It is used more or less as a 'sensitising concept' (see Glaser and Strauss, 1967), to guide the analysis in this study.

For this financial analysis we draw upon the ideas of Ricardo. His book 'The principles of political economy and taxation', first published in 1812, was the basis for an explanation of land prices as arising from scarcity, and the level of land prices as being a residual between the value of the product of the land and the costs of production. J.S. Mill (in 1849, in Principles of political economy) predicted from this that levying a tax on land would not affect the price of the product of land and would be borne by the land owner. And Henry George (in 1879, in Progress and poverty) popularised this into the politics of land taxation. Nowadays neoclassical economics has much more refined tools for analysing land prices and predicting the effects of instruments (for an overview of the neoclassical economic approach to land prices see Lipsey, 1966), but the theory of land price is still basically the same and its prediction still provides the justification for most land policy instruments. In particular, the theory allows the incidence of the costs and benefits of applying policy instruments to be predicted.

1.5   Influencing the residential environment

In the preceding sections, we have dealt with economic and institutional aspects of housing development. We now turn our attention to an important outcome of the housing development process: the residential environment. The residential environment is the combination of houses, plots, infrastructure, public spaces, facilities, and how all these elements are combined into one residential development, or housing scheme.

Decisions about the residential environment

During the housing development process, many decisions are taken which influence the residential environment, for example about the level of primary and secondary services, the housing density and the site layout. Local planning authorities try to achieve a certain minimum quality of residential environment. Therefore, the level of services and other aspects of that environment are partly determined by non-negotiable 'quality standards', based on social or safety considerations, fixed by the municipality or a higher level of government. Concerning other decisions about the residential environment, financial considerations play a role, and there may or may not be negotiation about these. For example, it is generally so that the developer of a site – whether it is a private or a public party – has to pay for the primary or on-site services (see Needham and Verhage, 1998).

Often there is a certain financial margin in a development process. Total income minus total costs (including expenditure on the fixed 'minimum level' of quality and a certain minimum profit to induce developers to supply) is then positive. Decisions about what to do with this financial margin are subject to negotiations. Decisions concerning the residential environment that are influenced during the housing development process are those about which there is a possibility for negotiations. The quality standards, fixed either by law or by practice, are not very interesting in this respect. They do not emerge from or during the housing development process, but are an input in this process from the outside. Although they have financial impacts, they are not reconsidered in the development process. Our focus is on decisions with a financial component that influence the residential environment, in as far as they can be influenced by the actors in the development process (see figure 1.1).

A few words about residential quality

The residential environment has a certain quality, and we assume that this quality is at least partly influenced during the housing development process. The quality of the residential environment can be defined as the extent to which the residential environment corresponds to the demands and preferences of residents, planners, developers, and politicians. All these different actors in the development process have their own ideas about quality. These ideas are influenced by the objectives and the background of the actor. It is not the aim of this study to determine what should be seen as residential quality. From other research (e.g. Kuiper Compagnons, 1990, 1991; Winter et al., 1993; Ministerie van VROM, 1996; Carmona, 1999), certain properties of the residential environment that influence its quality can be distinguished. Examples are: housing density (including proportion high rise/low rise), mix of tenure, amount of public open space (parks, play spaces), proportion green/water/hard surfaces/building plots, cycleway and footpath networks, design (including street furniture), facilities (schools, shops, clinics, ...). This study concentrates on the way in which during the housing development process decisions about expenditure on such aspects of the residential environment are made. In section 2.2, we return to the way in which this subject is dealt with.

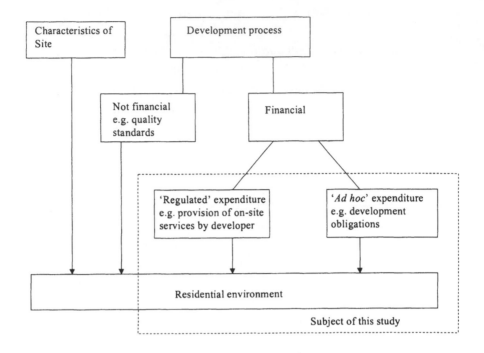

Figure 1.1   Influences on the residential environment

For this study, it is useful to distinguish between primary and secondary services. Primary services are essential for the new development and are provided within the boundaries of the site being developed and up to the boundaries of the building plots. These services include access roads, drainage, gas water and electricity, open spaces, and also the costs of making the detailed plans. They are sometimes called on-site services. Then there are secondary services which link the development site to existing infrastructure networks, and any other costs occasioned by the new development, such as school places, expansion of sewage purification plant, etc. They are usually provided outside the development site, hence the alternative name off-site services. The infrastructure within the boundaries of the building plot and which connects the building to the primary services, is the responsibility of the building developer.

In chapter two, we return to the way in which the analysis of the residential environment is dealt with methodologically. Here, we just recall that the focus of this study is on expenditure that is judged by the actors in the process as being aimed at increasing the residential quality. This broad interpretation is necessary because of the comparative character of the study. A list of what the researcher sees as residential quality is probably biased by the cultural background of the researcher. What he finds important in a housing scheme might not be seen as such by people from another background, or with another point of view.

1.6 Putting the pieces together

Blowers (1980: 110) remarks: 'Land allocation and development (...) is a product of the interaction between the market and the planning proces.' We agree with the observation that both the market and the public policy, or planning, determine the results of a development process. Therefore, to analyse the way in which the interactions lead to a product, we propose an analytical framework for the housing development process that combines a financial and an institutional analysis. For the institutional analysis, we have drawn from different authors, but the result can be seen as an elaboration of Healey's (1992b) 'institutional model of the development process' (see chapter four). For the financial analysis, a residual analysis of land prices is used (see chapter three). The analytical framework for this study attempts to link these two parts of the analysis. To that aim, a central assumption is posed, around which the analytical framework is constructed. That assumption is that a crucial phase in the development of new housing is the process of land conversion, and that by focussing on this process, the micro-economic and the institutional analysis can be linked.

A focus on the process of land conversion

Land assembly and land development are the activities in the housing development process where the two parts of our analysis – economic and institutional – come together. These activities together can be termed the process of land conversion. According to Barret and Healey (1985: 349): 'It is this process of change which interrelates all the other dimensions; for

21

example, change in land ownership involves establishing value as the basis for negotiating price, change in use may involve change in value, management and ownership and also development. The process of land conversion (whether upward or downward in value terms) forces consideration of the interaction effects of one dimension upon other'. As Barret and Healey suggest, we place the activities of land assembly and land development in the centre of the analysis of decision making in housing development processes. The argument for this is fourfold:

–   the residual land price calculation is explained in sections 3.1 and 3.2. Briefly, it implies that the price of land is derived from the price of what is produced on the land, in this case houses. The price that a developer is prepared to pay for the land is calculated by subtracting the costs of housing construction, of land development, and any other costs which might appear in the process of housing development (including a possible profit margin) from the income yielded by the sale of the houses. As a result, if the price that is paid for the land during the land assembly is exactly the maximum price thus calculated by the developer, there is no more room for a financial margin to occur in the development process. In other words, the prices paid in the first phases of the development process determine to a large extent whether or not a financial margin might appear in later phases of the process. However, if the market price of the end product (the houses) increases after the residual land price calculation has been made, a financial margin can appear in a later phase of the process only;

–   the party that undertakes land assembly and land development makes an investment that brings a high risk. Often, the party that acquires the land does not itself make the decision whether the site is going to be developed, and thus will increase in value. This means that by buying the land, this party takes a risk which it will want to see compensated by high possible returns. The housing construction itself, once the land has been acquired and serviced, is not very different from a 'normal' production process of durable goods. Investment bears fruit in a short term, the risks are not especially high. Therefore, the gains on housing construction do not play as important a role as the gains on land development;

22

- as a result of land use regulation through the planning system, land is a scarce resource for housing development. It is not possible to buy any piece of land and then develop houses on it. This makes the land a crucial variable in the development process. The party who owns the land where development is going to take place has – purely because of the fact that it owns the land – an influential position. If it does not cooperate, the process might not be carried through;
- an additional reason for the important place that is accorded to the initial phases of the development process is that decisions that are taken here set the conditions under which the development has to be carried out. The characteristics of a housing scheme are to a large extent determined during the land development. Parcellation and infrastructure provision set   limits for what is possible during the housing construction.

For these reasons, this study concentrates on the land assembly and land development. In these activities, the basis for decisions about expenditure on the residential environment can be found. This corresponds with the work of Dieterich, Dransfeld and Voß (1993b), for the German *Ministerium für Landesplanung und Raumordnung* (Ministry for regional and land use planning). In a comparative analysis of the land and property markets in five European countries, they distinguish five different types of development processes. According to Dieterich et al. (1993b: 141-142), the essential distinction between different development processes is who has the ownership of the land during the activity of land development? This is essential because it determines who is responsible for the servicing costs and who receives a possible financial margin. In the first three types of development processes, development is carried out by one temporary land owner. The first type is temporary ownership by the municipality. The second type is characterised by temporary ownership by a public or a public/private body. In the third type, a private developer temporarily owns the land. In the next two types of land development process, nobody acquires all the land within the area to be developed. Ownership of the plots is divided between the original owners. All those separate landowners individually will not be able to provide the necessary services.

A public body will have to do this, and it can do it either with (type four) or without (type five) using public powers.

It is possible to distinguish another type of development process that is not distiguished as such by Dieterich et al. (1993b). When the land development is carried out by a private developer, there are two possibilities. It can be the private developer's aim, after developing the land, also to construct the houses on the site. The private developer then takes care of the realisation of the entire housing scheme. However, it also occurs that a private developer does not realise the housing scheme, but only carries out the land development. Once that is finished he sells serviced building plots to the house builders. In this study we will see that this type of development process has some features distinct from what is described by Dieterich et al. as a 'type three' development. This type is considered as a variant of the land development by a private party as temporary landowner.

The actor that receives the financial margin

Following Needham and Verhage (1998) these five types of development process and their consequences for the incidence of servicing costs and the financial margin are described below.

–   Type one: temporary ownership by the municipality. In this case, if the municipality bought the undeveloped land at its existing use value and if it sells building plots at market value, then the municipality receives the financial margin. If the municipality bought at a price above existing use value, then the financial margin is shared between the first owner and the municipality. We can also say more about who pays the costs of the servicing. It is not possible to raise disposal prices of building plots by loading costs for servicing onto them. If those costs are charged to the land development process, they will be included in the price of the plots and therefore reduce the financial margin.
–   Type two: temporary ownership by a public or a public/private body. What applies to the first type of development process applies equally to this second type. The financial margin is received by the body

responsible for the land development, or is shared between that body and the private owners. The public/private body bears the costs of the primary servicing. The costs of secondary services can be included in the price of the plots. They can also be borne by the taxpayer. In the latter case, the remaining development gain will be higher.

- Type three: a private developer temporarily owns the land. This private owner either develops only the land, or he develops houses on it also. In both cases, the process is rather similar to the first two types, except that it is a private body which receives the financial margin, possibly sharing it with the first owner of the land. If there are no special arrangements, the costs of secondary services are borne by the taxpayer. However, the costs of secondary services might be charged to the private developer by means of development obligations. In that case, the remaining financial margin is reduced.

- Type four: no temporary landowner, no public powers used for land development. In this case, which actor receives which part of the financial margin depends on private agreements made with the local planning authority. However, with a voluntary agreement it is unlikely that the landowners will be prepared to pay more than the cost of the primary services. So they will receive the full financial margin. Taxpayers will bear the costs of the secondary services that will have to be carried out by the local planning authority. If it is thought incorrect or unacceptable that the land owners should receive the entire financial margin, then there might be taxes or levies on development gain.

- Type five: no temporary landowner, public powers used for land development. Here, the outcomes depend upon the contents of the public powers which are used. If these allow all the costs of the services to be recouped from the land owners, then the latter still receive the (reduced) financial margin. A different kind of public power is possible whereby a charge is levied equal to the servicing costs plus the development gains. In that case, the costs of services are paid out of the levy, the rest of the levy (which corresponds with the financial margin) goes into the public purse, and the landowner realises no more than existing use value.

The types of development process can help to explain the use to which a possible financial margin is put. In each type of process, a different participant or combination of participants receives the value increase of the land, according to its role. Each participant has different objectives. These objectives determine the use to which it wants to put a possible financial margin, and thus the possibilities for expenditure on the residential environment. In the next section, we focus on the actors in the development process, and the way in which they influence, in their interactions, the use of the financial margin.

Decisions about expenditure on the residential environment

Above, the question of how the residential environment is influenced during the housing development process has been translated into the question of how, during the development process, decisions about expenditure on the residential environment are taken. The first part of a response to this question – where does the money come from? – has been dealt with in what we have called a financial analysis of the development process. The second part – what determines the use of the money – asks for an institutional analysis of the same process. This requires considering the following:

–   the activities, actors, roles, and objectives;
–   the interactions;
–   the power balance.

What we want to know is which actor was responsible for land assembly and land development. Here lies the link between our two types of analysis. In the preceding section it has been shown how this central actor influences a possible financial margin in the housing development process. The question that concerns us here is how decisions about the use of this financial margin are made. To understand this, we must concentrate on the interactions in the process. These can be understood as follows.

All the actors in a development process have certain objectives they want to realise. Because there is interdependence between the actors and no single actor is in charge, none of the participants separately can realise

26

anything, the participants start to interact. These interactions can take many forms, e.g. negotiations, economic transactions, legal supervision. In these interactions, the actors use the power they have to obtain their objectives. The power/dependence that is expressed in interactions can depend on the actor (e.g. a public actor has public law as a basis for power, whereas a private actor does not), and on the role an actor takes on (e.g. when they are landowner, both public and private actors have a crucial resource as a basis for economic power over the other actors). In combination with the objectives of the actors, the relations of power and dependence – as they are expressed in the interactions – can be used to explain how and why decisions in the housing development process are taken.

This study focuses on these decisions as they relate to expenditure on the residential environment in a broad sense: we are concerned with all decisions that are judged by the actors in the process to affect the residential environment. A combination of the two types of analysis – financial and institutional – is used to interpret these decisions. Dealing with the relation between housing development process and residential environment in this way allows for a comparison of processes that are very dissimilar at first sight.

Combining an institutional and an economic perspective

We would like to find out how financial aspects are incorporated in the institutional considerations, and conversely how institutional considerations influence the financial course of the process. We trace this link by focussing on the decisions that are taken during housing development processes, concentrating on decisions concerning the residential environment. Both financial and institutional considerations influence these decisions. Therefore, by studying how the decisions in a housing development process are made, and by uncovering the considerations that lie behind them, we trace the link between the two driving forces that we have distinguished in the process of housing development.

The questions we are concerned with can be described as dealing with the coordination of economic activities within an institutional context.

Since Coase (1937) investigated the question why not all economic exchange was carried out through markets, several strands of economic theory have been developed in this field. The field of study of these 'institutional economics' is clearly depicted by Douma and Schreuder (1991). They give an overview of different economic approaches to organisations. Thompson et al. (1991) deal with the same kind of questions, but put them in a broader context when they speak of the coordination of social activities. Without going deeply into institutional economic theory, we use some concepts developed there to present the link between the constituent parts of our investigation.

Both Douma and Schreuder, and Thompson et al. start from an economic perspective. As such, they are concerned with transactions between people. More specifically, they investigate how such transactions are coordinated. The importance of coordination of (economic) activities is well pointed out by Thompson et al.: 'Various agents and agencies can be 'ordered', 'balanced', 'brought into equilibrium', and the like, by the act of coordination. Without coordination these agents might all have different and potentially conflicting objectives resulting in chaos and inefficiency' (1991: 3). They observe that the classical economic view on how such transactions are coordinated – i.e. by the concept of the market – is not always applicable. A second coordinating principle is introduced which is termed 'hierarchy'. From their slightly more restricted point of view, Douma and Schreuder speak in this respect of 'organisation'.

To complicate the matter further, a third coordinating principle is described by Thompson et al., i.e. the network. 'If it is price competition that is the central coordinating mechanism of the market and administrative orders that of hierarchy, then it is trust and cooperation that centrally articulates networks' (Thompson et al., 1991: 15). In housing development processes, the question of coordination is central. In chapter three, the emphasis is put on the role played by the market as a mechanism of coordination. In chapter four, the more 'organisational' or 'hierarchical' – we call them institutional – aspects of coordination are investigated. In chapter five, both angles are combined by a focus on decision making about the residential environment. In that chapter, we will see that housing development always contains elements of markets and of hierarchies, in various proportions, but that neither of the two on its own can explain the course and the outcomes of the development process. The network is required as an additional coordinating principle to describe and understand housing development processes. In the case studies, we will be able to observe all three types of coordination. We return to them in the concluding chapter six of this book, to draw together the threads that have been spun out in the other chapters that are each on their own a partial analysis of the relation between local policy for housing development and the residential environment.

2 Cross-national comparison of housing development processes

2.1 An approach to policy analysis

In chapter one we developed our view of the process of housing development. The next step in our study is to determine how to study this process. That is the subject of the present chapter. We explain the choice of the format for the investigation, and the research methods. To that aim, the nature of the study subject, the design of the study, the wish to carry it out from a cross-national perspective, and the type and scope of conclusions that we want to draw are addressed. In other words: how has this study been carried out, and why?

Each actor has his own process

Our perspective on policy making, as it occurs in housing development processes, is well summarised by Healey: 'Public policy, and hence planning, are (...) social processes through which ways of thinking, ways of valuing and ways of acting are actively constructed by participants' (1997: 29). Each of the actors in the process has his own construction of it. If asked to do so, he will describe the process in a different way. The picture that the researcher forms of the process is a construction too, informed by the view of the different actors. This activity of interpretation or construction is referred to by the notion of hermeneutics (Guba, 1990). In this respect, the term 'double hermeneutics' is used to point at the fact that the inquirer makes an interpretation of the interpretations of others. After all, the texts or the interviews by which the inquirer gets his data are interpretations, or 'constructions' made by others. The notion of dialectics points at the interactions between inquirer and inquired. The way in which the inquirer understands the inquired (his construction) is constantly shaped and refined by confrontations with the studied phenomenon.

If each actor in a development process has his own construction, and for his reconstruction the researcher can use only these reconstructions, then what is the task of the researcher? To explain our position, we use the notion of positivism, and the critique of this when applied in the social sciences. In the words of Guba, the basic ontological assumption of positivism is that 'There exists a reality out there, driven by immutable natural laws. The business of science is to discover the 'true' nature of reality and how it 'truly' works' (1990: 19). The way in which this should be done is by looking objectively at the phenomenon to be studied. However, it is admitted that the inquirer can be biased. To counteract this possible bias, the inquirer should use a scientific methodology that excludes influence from the inquirer on the observed phenomenon. This typically takes the form of '... hypotheses (...) stated in advance in propositional form and subjected to empirical tests (falsification) under carefully controlled conditions' (Guba, 1990: 20).

There are a number of arguments why this perspective on science is challenged (see Guba, 1990: 25-26). They can be summarised in the basic assumption of the social constructivist paradigm, i.e. that objectivity is not possible. Reality is not 'out there', but is the result of an active process of inquiry, in which the inquirer's perspective is selective and theory and value laden. As a result, when different inquirers study the same phenomenon, they will come up with different results. This has important consequences for methodology. Falsifying hypotheses formulated in advance does not comply with the idea that the observations that an inquirer makes are determined by the chosen perspective (the theory). Researchers therefore have to find other ways to make sense of the phenomena they are studying. They do this by using a hermeneutic and dialectic approach to the studied subject. Guba works this out in a social constructivist perspective, explained as opposed to a positivist perspective. We choose a different approach, for which we draw upon the work of Galtung.

In his book 'Methodology and Ideology' (1977), Galtung explores the relation between the social scientist and the society, and the way in which this shapes the theory and methods of social research. He critisises the 'traditional scientific activity', and proposes an alternative perspective. However, he does not oppose this alternative perspective to the positivist perspective. He presents it as a modification. As such, in Guba's (1990) view, Galtung's approach can be categorised as post-positivism. The alternative proposed by Galtung is a trilateral scientific activity (1977: 41-71). He describes this using the notions of theory, data, and values. With the use of these three notions, Galtung draws up what he calls 'the science triangle'. On the top angle he puts data, on one of the two base angles theory, and on the other value. The side that connects data with value is called criticism, and the side that connects theory with value is constructivism. The side that connects data with theory is termed empiricism. This empiricism is very clearly an approach that Guba (1990) would fit into the positivist paradigm. We take it as representing the positivist scientific inquiry.

This triangle gives us a very useful tool to describe our position. Hopefully Galtung will not object if we summarise his much more refined analysis and the conclusions he draws from it as follows. Both constructivism and empiricism deal with only two of the three notions mentioned above. Empiricist scientific inquiry does not allow values to play a role in the analysis and is therefore concerned with the relationship between theory and data. It aims at formulating statements that can be either true or false. This is a partial approach because of the exclusion of values. In our study of development processes where several actors with different values interact, the notions of true and false are clearly influenced by values (see the quote from Healey at the beginning of this section).

Constructivist scientific inquirey is concerned only with the relationship between theory and values, and excludes the idea of data representing reality. Its aim is to formulate statements that can be desoribed as adequate or inadequate. In our view, this too can only give a partial description of our subject. For our interpretation of what happened in the studied development processes, we draw on the reconstructions of the different actors. But these are confronted with empirical data, such as money flows, time schedules in which activities are carried out, and roles that different actors play. Thus we avoid following the reconstruction of one or two dominant actors. It is a verification that the researcher needs to validate his reconstruction of the development process.

The third type of scientific inquiry distinguished by Galtung, criticism, is concerned with the relationship between values and data. The conclusion of this type of inquiry is in terms of good and bad. Guba (1990) uses the term 'ideologically oriented' to depict this branch of inquiry. The different approaches that can be put under this heading are characterised by a rejection of the claim of objectivity made by positivists. Their solution is deliberately to choose a value position. Again this gives a partial approach to the study subject because this choice for one perspective excludes valuable insights that could be derived from other possible constructions.

When trying to make sense of the practice we see around us, Galtung argues, theory as well as values as well as data must be taken into consideration. He terms this approach a trilateral science. It aims at formulating statements that can either work, or not work in practice. This idea about scientific activity lies at the basis of the analysis in this book. It has consequences for the kind of conclusions that can be drawn: we return to that below. First we take a closer look at the specific form in which this study is carried out, i.e. as a cross national comparison.

A cross-national comparison

The choice to carry out a study cross-nationally is often considered to bring along a whole range of methodological problems. These problems, however, are not fundamentally different from any other comparative (sociological) research. Or, as Øyen puts it 'All the eternal and unsolved problems inherent in sociological research are unfolded when engaging in cross-national studies' (1990: 1). That observation does imply that the researcher has to solve a whole range of methodological problems. These problems are related to the differing contexts of the cases, differences that are bigger when the cases are chosen in different countries, each with their own habits and traditions. Why then, a cross-national perspective? For certain purposes, a cross-national comparison can be more interesting than a study within one country.

The reason for carrying out this study cross-nationally is that local government in different countries operates within certain country-specific limitations (policy of higher tiers of government, legislation, planning practice, structure of the development industry). Most of these factors will be similar within one country. Moreover, policy choices are constrained within the knowledge of alternatives, and within one country there are usually ways which are 'taken for granted'. For those reasons, a study of the relationship between local policy for new housing development and the residential environment can better be carried out cross-nationally. That exposes implicit assumptions within one country and shows that more policy options might be available.

The different countries are not the subject of the study but the context of the study subject, i.e. the housing development process. The aim is not to compare different countries, but to compare different housing development processes. This can be put in more methodological terms. The process of housing development can be described as the independent variable in our study. We want to find out whether variations in this independent variable lead to different outcomes, that is, to differences in the residential environment which is the dependent variable. The reason to carry out this study cross-nationally is to have more variation in the independent variable (different forms of housing development processes). The purpose of this is to construct a framework that has a large scope, i.e. that is applicable in a variety of situations. The variety in contexts in which the analytical framework is applied is also a means to investigate the interference of 'context variables'. If a certain assumed relationship is observable in a number of different situations, this is an argument in favour of the existence of the assumed relation: it is then less likely that it can be attributed to an intervening variable, because in all cases, different variables intervene. In methodological terms, the approach that is chosen in this study can be described as a 'most different systems design' for comparing cases (see Korsten et al., 1995).

Type and scope of conclusions

The approach that is chosen has consequences for the type and scope of the conclusions that can be drawn. As explained in chapter one, we want to find out how interactions in the process of housing development – and hence the outcomes – are shaped by context variables, in particular financial and institutional aspects. In the words of Teisman, we '... study the way in which the strategy of the actors is influenced by characteristics of the context, and how in turn this influences the interactions between the actors. These interactions are never one-way. The strategy of the actors is not determined by the context, after which the strategies determine the interactions' (1998: 91 – translation RV). As a result of this continuing adaptation of the actor's strategies to the context, predicting the outcomes of one (future) housing development process is not possible on the basis of the study of a number of other housing development processes.

What our study can show is which elements play a role in determining the outcomes of housing development processes. This allows us to reconstruct afterwards why in a particular case a particular outcome was realised. The aim is to interpret this in such a way that it can serve to formulate statements (a 'theory') that work, that is, that generate useful insights into the relation between the housing development process and the residential environment. Thus, it can help the actors involved in the development process to reflect upon their actions. To use the words of Teisman again: 'Each new process offers its own opportunities and threats, which the actors involved can discover upon critical reflection' (1998: 90 – translation RV). The aim of the framework is to guide this reflection by offering the actors in the process a window through which to interpret their actions. These actions can of course also be interpreted in other ways, which would probably teach the policy makers other things. That raises the question why this study has come up with the framework that it has. The following section deals with that.

2.2   A framework for a cross-national comparison

With the above perspective on the study of social phenomena, we encounter a dilemma which is also acknowledged by Galtung. He formulates it as follows: 'On the one hand we want to benefit from the advantage of a well-crafted theory as the great organiser of known and unknown insights. On the other hand, we do not want to mistake theory for reality. We want the theory to be our tool, not vice versa' (1990: 100). Galtung's solution to this dilemma is a plea for eclecticism. Each new, well considered reconstruction of a social phenomenon adds to our understanding of it. Galtung's description of theory formation in social sciences is interesting because he points to the value of a multitude of perspectives in social research. He acknowledges the fact that each perspective, or theory as he calls it, is a construction and necessarily partial. This underlines the importance of different constructions or frameworks in social research, but at the same time acknowledges the relative value of each framework.

The necessity of a framework

When engaging in international comparative studies – but as we have said above, this is true for any comparative study – one inevitably encounters differences between the elements of comparison, in our case the processes of housing development. Basically, these differences can be approached in two ways. One is to turn the attention to these differences and use them to illustrate the importance of culture, practice, legislation and whatever is seen as the reason for the differences between the countries. This approach leads to conclusions stressing the uniqueness of each country. It also leads to critical remarks as to the use of international comparative research, because why should we study other countries when we cannot really learn from them because we cannot copy their practices? After all, these practices are a result of their culture, customs, legislation, or whatever. The other approach to cross-national comparisons is to concentrate on the similarities between countries, rather than stressing the differences. In such an approach, the different countries are seen as different contexts in which people do similar things. The fact that they do these things in different ways makes the comparison interesting, because it enables people to reflect upon the way they usually do things. It can show alternative ways of acting and enables us to expose implicit assumptions and to question habits or practices that are usually 'taken for granted'. The second approach is chosen for this comparative study.

For such an approach, an analytical framework is required that allows phenomena in different countries to be compared. For this study, we need a framework for studying housing development processes which can be applied in several countries. At first sight, the differences between these processes are obvious. In much the same way, when we walk around cities in different Western European countries, the differences between what we see and what we are used to seeing stand out. With regard to the residential environment, Van Der Krabben gives an example of such differences and their relation to the prevailing mode of housing development: '... it is remarkable that in the Dutch housing market dwellings are usually built in large quantities by property developers, while in other countries – Germany and Belgium are perhaps the extreme examples in this respect – many more owners build their own dwellings in more or less isolated locations. The consequences for urban spatial structures may be far more drastic than generally seems to be assumed' (1995: 35).

However, without trivialising these differences, we can also argue that we see much the same. Although the details are different, we are walking on a sidewalk. We are looking at buildings situated in gardens. Along the road are parking places, and if we look up we see street lighting. Although we may tend to focus on the differences, a lot of things we see are the same as they are in our own country. To notice these similarities we need a certain level of abstraction. Although houses in the Netherlands and in Germany may have very different features, they can both be considered as a house and as such can be compared. The notion of 'house' provides a 'framework' for a comparison of buildings that may be very different when one does not qualify them as houses to begin with. In much the same way, for our study of planning policy and legislation on a local level, we want to use a framework that allows us to compare, rather than only to apreciate the differences between the studied countries. We construct this framework around the notion of the housing development process. It is assumed that this exists in all countries. That is, in all countries activities, actors, and interactions occur when a green field is transformed into a residential area. The aim of this study is to construct, apply, and refine a framework for the study of housing development processes in different countries.

In chapter one, the outlines of the framework are presented. In the following chapters it is further elaborated. By defining in advance the elements of the process that we turn our attention to, we limit our description of what happens. This is the most noticeable in the diagrammatic representations of the development process that are used throughout the study. These diagrams show only selected facets of the housing development process. Moreover, policy processes never fit snugly into such diagrammatic representations. Presenting these processes in such a way always does harm to the nuances that can be observed. But reconstructing complex processes in such a way that we can make sense of them requires a selective viewpoint and a certain level of abstraction. It is the task of the researcher to argue why exactly these facets are shown, and what they can teach us.

Analysing and comparing residential environments

As described in chapter one, the reason why housing development processes are studied is to find out how, during this process, the residential environment is influenced. The way in which this question is approached requires some elaboration. It is our contention that comparing 'residential environments' in different countries, to say which is 'the best' or has 'the highest quality' makes no sense. The perceived quality of the residential environment is to a great extent determined by the context. Depending on people's culture, interests, social situation, and a lot of other factors, the residential environment is judged differently. Or in the terms used in this chapter, different constructions of residential quality exist. Of course, at the level of social or safety standards, there is a broad, intersubjective consensus on a certain minimum quality. But, as argued in section 1.5, that is not the primary interest of this study. Moreover, in the type of cases we have chosen (i.e. new greenfield development in Western-European countries) the level of regulation is so high that this minimum level is always reached. What we are interested in is how within one case the residential environment is influenced by aspects of the development process.

To find that out, we need a way to describe the residential environment. As shown in figure 1.1, from our perspective it is not relevant to describe how specific characteristics of the site lead to a specific residential environment. That influence goes beyond the development process. We want to focus on aspects of the residential environment that are influenced during the development process. Thus, it becomes

39

interesting when during the process the actors manage to use site specific characteristics in an unforeseen way, influencing for example the urban design of the area. We also want to describe the residential environment in terms that allow an appreciation of its quality. Although we do not make that value judgement, we are interested in finding out how aspects that are broadly considered to determine – at least partly – the quality of the residential environment, are influenced during the development process.

For these reasons, the decision making about the residential environment is described as the decision making about a selected number of aspects of the residential environment. The selection of these aspects is based upon studies into the quality of the residential environment (Kuiper Compagnons, 1990, 1991; Winter et al., 1993; Ministerie van VROM, 1996; Carmona, 1999). These studies are used to find characteristics that are considered to influence the quality of the residential environment, and that are influenced during the development process. With the use of these criteria, the following aspects have been selected:

–   housing density;
–   urban design;
–   public facilities;
–   mix social/market sector dwellings.

Although the quality of the residential environment is not judged in terms of these aspects, by selecting a limited number of aspects of the residential environment we do take a position as to what is considered as residential quality. Nevertheless, the characteristics that have been selected to represent the residential environment are neutral. Density can be high or low, urban design can be well thought out or neglected, public facilities can be absent or present, the housing scheme can be an equal mix of social and market sector dwellings or only one of the two sectors can be present. This study aims at describing and understanding how during the development process decisions on these variables are taken. The objective is to understand how all the parties involved realise together a certain residential environment. This should give insight into how objectives for good residential quality can be reached. But it is left to the actors in the process to determine these objectives.

2.3   About the research design

So far, we have mainly been concerned with the theoretical notions underlying the research reported in this book. In this section, we turn our attention towards the empirical part of the research. We consider the way in which the empirical data is gathered, and the way in which it is used.

A case study approach

The reason why the choice for a case study design is made is very well described by Huberts and De Vries: 'Interpretative researchers assume that people base their actions on their interpretation of significances. Policy processes must be understood and analysed from the perspective of these interpretations. The researcher has to empathise with the actors and if possible participate in the process. Qualitative methods are used to formulate statements that are valid for the studied cases. Case studies are central in this approach' (1995: 58 – translation RV).

We acknowledge that to a certain extent each housing development process is unique, because of its particular constellation of actors in a particular context. Nevertheless, we want to enable a better understanding of the process. This asks for an in depth study of a variety of housing development processes which can serve as 'substitute experiences', a notion borrowed from Abma (1996). Several case studies are carried out, instead of only one, because a multiple case study design opens the way for comparison. Experiences from one case study are taken along into the next. Thus, the method enables us to learn and at the same time to put into perspective the things that have been learnt. Also, studying different cases allows constant interaction between theoretical considerations and empirical findings. Insights from one case can be applied in another to see their usefulness. Thus, the researcher goes – in a more intense way – through the same learning process as the reader of this book.

As a result of this process, and as a result of the chosen methods of data gathering, the level of information between the cases varies. Experience from former cases has sometimes led to new questions in subsequent cases. We have sometimes taken the liberty to ask other questions in different cases. In that way, a focus on interesting elements of

41

each case was possible. In the 'case study files', which can be found throughout chapters two to five of this book, the cases are represented in the same chronological order as in which they were investigated, to allow the reader to trace such changes. The case study files are structured as follows:

- chronological overview of events (section 2.4);
- financial analysis (section 3.2);
- actors, roles and activities (section 4.2);
- relations of power (section 4.5);
- interactions (section 5.2).

We have used the term 'substitute experience' to describe the role of the cases. For the case descriptions to offer such an experience, a presentation at two levels has been chosen. In the case study files, relevant data from the cases are presented in a descriptive way. At another level, these case descriptions are used in the main text of the book to elaborate and illustrate the chosen theoretical perspectives. The theoretical notions and the descriptions of the way in which they can be observed and used in the case studies are as much as possible developed parallel to each other.

A characteristic of case studies which is often observed is the possibility of data triangulation. This refers to the use of different information sources and different methods of analysing the information from these sources, which allows the findings from one source to be cross checked with the findings from another source. This approach is used in this study. Data about the cases is gathered in interviews with representatives of different actors involved (e.g. planning authorities, private developers, housing corporations), by the analysis of documents (such as local plans, maps, agreements, financial accounts, commercial brochures, newspaper articles), and from site visits. For almost all of these data sources, the researcher depends on the willingness of the actors involved to cooperate and to provide information. It has not been possible in each case to get answers to all the questions. This is regrettable, but unavoidable in this kind of research. The financial analysis has been affected most by this problem. It was not possible to obtain all the required information in all cases due to the sensitivity of ongoing negotiations.

42

To make sense of the abundance of data encountered when engaging in a case study, the choice is made to structure it beforehand. The cases are approached with the use of an analytical framework based upon work of others who have studied development processes (e.g. Gore and Nicholson, 1991; Monk et al., 1991; Eve, 1992; Healey et al., 1995; Dieterich et al., 1993b; Barlow and Duncan, 1994; Lacaze, 1995; Van Der Krabben, 1995). From these studies, it becomes apparent that housing development can be approached in numerous ways, each of them providing different, often supplementary insights into the process. Although there are a large number of variants, the approaches towards housing development can be seperated into the ones with a focus on the economic processes (e.g. Monk et al., 1991; Eve, 1992; Barlow and Duncan, 1994), the ones with a focus on institutional processes (e.g. Healey et al., 1995) and the ones trying to combine both focuses (e.g. Dieterich et al., 1993b; Van Der Krabben, 1995). This observation has shaped the form of this study. We choose to apply two different approaches to the process of housing development, i.e. a micro-economic approach, and an institutional approach. These approaches are combined in such a way as to show cross-linkages between them. Thus, we expect to be able to paint a more complete picture of the housing development process.

Let us illustrate this with an example. A micro-economic view on housing development focuses on prices and money flows and the way in which these influence the actors' behaviour. Thus, it might explain why at a certain time certain houses are built (i.e. as a logical result of supply and demand interfering). However, it is widely acknowledged in economic studies that housing development takes place in a market that is especially influenced by institutions. For example, central or local government regulation influences supply: what, where, and when houses can be built. Demand is influenced by the tax system and by the possibility to obtain mortgages, but also by cultural factors, such as what type of houses people want to buy, and where they prefer to live. Such factors can be incorporated into an economic approach as factors which influence supply and demand and how they interact. Housing development processes can also be approached taking these institutional factors as a starting point.

This involves a focus on actors and strategies to see how these are influenced by the institutions.

It is our contention that each of the viewpoints on its own provides useful insights. With Galtung, we would like to argue that 'A good theory, then, would be like a family or combination of a number of classical theories, squeezing each one of them as far as possible for their intellectual content, neither believing, nor disbelieving completely in any one of them' (1990: 101). However, trying to amalgamate both viewpoints into one framework is seen as a challenge. To that aim, the insights generated by each of the two complementary approaches are combined to provide a coherent theory about housing development. We value the eclecticism that this study thus inevitably involves.

The choice of the countries and the cases

We focus on the development of housing on greenfield sites. There are three reasons for limiting ourselves to such sites. The first is that brownfield development often requires subsidies from higher levels of government. As the focus of this study is on the influence of local policy for housing development, we want to avoid as much as possible involving any other than the local level of government in the analysis. The second reason lies in the comparative character of the study. A financial analysis is an important part of the study. Many more variables influence the money flows and the outcomes of brownfield development than of greenfield development. (Financial) analyses of brownfield development in different countries are therefore likely to be more difficult to compare than for greenfield development. The third reason is that one of the occasions to carry out this study, the changing situation in the Dutch land market, concerns the development of housing on greenfield sites.

As explained in chapter one, we distinguish different types of development processes according to who (temporarily) owns the land during the process of land conversion. Dieterich et al. (1993b) distinguish on the basis of this the following five different types of development processes (see section 1.6):

- temporary ownership by the municipality;
- temporary ownership by a public-private body;
- temporary ownership by a private developer;
- no temporary ownership (first landowners retain their land), no use of public powers;
- no temporary ownership (first landowners retain their land), using public powers.

On the basis of our investigation, the category of temporary ownership by a private developer must be divided in two. The case in which the private developer is a house builder, who after the land conversion also builds the houses is distinct from the case in which a private developer only does the land conversion, and then sells the serviced building plots to house builders. Thus, we distinguish between six types of housing development processes. In the chosen cases, the six different types of development process all occur.

The countries in which the cases have been selected are the United Kingdom, Germany, France, and the Netherlands. Theoretical reasons for this selection are that the countries have a comparable level of socio-economic development, and that together they represent a wide range of governmental systems and planning systems (see, for example, Newman and Thornley, 1996). Besides, there are practical reasons. At the start of this study, the level of knowledge of land policy and housing development in these countries was such that it could be expected that the objectives of the present study could be reached in the time available. The same argument was used to limit the number of countries to four, one of which had to be the Netherlands because of the changes in Dutch land policy and housing development which were an important reason for this study.

Much of the knowledge of the countries concerned was collected in a study initiated by the German *Bundesministerium für Raumordnung, Bauwesen und Städtebau*, reported in the book series European Urban Land and Property Markets (1993, 1994). The end of that study project, marked by the publication of the report by Dieterich et al. (1993b), formed a starting point for this study. The network of contacts created for the study could directly be used for this study. These reasons for the choice of countries could be seen as compromises between theoretical and practical considerations. With Øyen, we would like to argue that 'Such compromises form part of the research process, but can at the same time yield windfall solutions, as the familiarity with a country provides additional information, increasing the value of the explanatory statements' (1990: 11). In fact, the choice of the countries might not be important. What counts is the cases that have been studied in these countries. The reason to study different countries was to be able to choose a wide range of different types of housing development processes as cases.

Nor is the choice of the cases themselves based only on theoretical considerations. Lots of different cases that all fit into the theoretical argumentation could be studied. Out of this large number, only very few have been selected. The cases should allow to gain more insight into how strategies of actors are influenced by the context, how they are expressed in interactions, and how this influences the decision making. To reach that aim, it is interesting to study a variety of cases, for it allows more varied 'substitute experiences'. This should teach us more about these subjects than if the study were aimed at very similar cases. If certain patterns recur in varied contexts, it is more likely that they occur often in practice. In any case, as Mitchel puts it, inferences from case studies depend on 'the validity of the analysis rather than the representativeness of the events' (1983: 190). Throughout this book, we try to demonstrate the validity of our analysis.

By choosing two cases in each of the four countries, it has been possible to have cases of each of the different types of housing development described above. In the next section, this chapter is concluded with a description of each of the chosen cases:

- temporary ownership by the municipality: Zwolle Oldenelerbroek and Arnhem Rijkerswoerd phase two in the Netherlands;
- temporary ownership by a public-private body: Rennes la Poterie in France;
- temporary ownership by a private house builder: Bishop's Cleeve and Cramlington North East Sector in the United Kingdom;
- temporary ownership by a private land developer: Bois-Guillaume les Portes de la Forêt in France;

- no temporary ownership (first land owners retain their land), no use of public powers: Stuttgart Hausen-Fasanengarten in Germany;
- no temporary ownership (first land owners retain their land), using public powers: Bonn Ippendorf in Germany.

1. Arnhem Rijkerswocrd
2. Zwolle Oldenelerbroek
3. Bishop's Cleeve
4. Cramlington north East Sector
5. Bonn Ippendorf
6. Stuttgart Hausen – Fasanengarten
7. Bois – Guillaume Portes de la Forêt
8. Rennes la Poterie

Figure 2.1   Situation of the cases

2.4   Case study files: chronological overview

Arnhem Rijkerswoerd phase two

Table 2.1   Core data Arnhem Rijkerswoerd phase two

Location	The Netherlands, Provincie Gelderland, near Arnhem	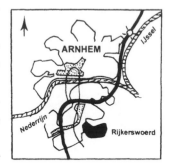
Size (ha)	40	
Number of dwellings	1429	
Housing density	35.7 dwellings/hectare	
Time period of development	1989 – 1994	
Initial ownership structure	Municipality owned the land at the start of the process	
Land developer	Municipality	
House builder	Private house builders, housing corporations	

Arnhem is a city with about 130,000 inhabitants, in the east of the Netherlands on the banks of the river Rhine. It is the principal centre of a conurbation of some 250,000 inhabitants. The city's most recent extension is called Rijkerswoerd. The development of this area of 150 hectares started at the beginning of the 1980s. We focus on the second phase of the development: 1429 dwellings and their surroundings, realised between 1989 and 1994. The whole area is developed in the 'traditional' Dutch way: the municipality has bought and developed the land and then sold serviced building plots to private developers and housing corporations. The municipality is responsible for the land development. The private and social housing developers were responsible for the design and the construction of the houses, within limits set by the municipality.

The plans for the realisation of a new residential area to the south of Arnhem started in 1975. In the first plan that was presented, the municipality envisaged the realisation of 3500 dwellings in a large area, with a large amount of green space. These plans became untenable when the municipality realised that it had to apply for *locatiesubsidie* (location

48

subsidy) to the central government to be able to finance the development. An application for this *locatiesubsidie* was successful only if the development corresponded to a set of norms regarding housing density, area of green space, plot sizes, etc. For Rijkerswoerd this meant that for the same area, 7000 instead of 3500 dwellings had to be realised.

At that time (in 1982), the central government had just presented its policy of 'bundled deconcentration' which directed the growth of big cities towards neighbouring smaller towns. For the city of Arnhem this meant that the neighbouring municipalities of Duiven and Westervoort were to provide housing for the people from Arnhem. This severely thwarted Arnhem's expansion plans. The municipality of Arnhem communicated its concern to central government and started negotiations with the municipalities of Duiven and Westervoort, and with provincial and central government. These resulted in an agreement in which the houses to be built were divided fifty-fifty over Arnhem and the growth towns of Duiven and Westervoort.

Thus, halfway through the 1980s, the municipality was allowed to build 4600 dwellings in the area of Rijkerswoerd.

Because of the uncertainty about the number of houses it would be allowed to develop, the municipality meanwhile drew up a plan that proposed a realisation of Rijkerswoerd in three phases. This division in phases should allow a flexible realisation of the urban extension. The first phase consisted of some 1000 dwellings, in direct relation with the existing city and with existing services. In the second phase, another 2500 dwellings were to be built, with some basic services like a park and a primary school. The development of 1000 more dwellings in phase three would allow the municipality to realise all the necessary services (especially a small shopping centre and another school) that would make the area a full, more or less independent quarter in Arnhem.

This plan, as laid down in 1985 in the *globaal bestemmingsplan* ('global' local land use plan) has been realised, with some modifications as to the number of dwellings be realised in each phase. The case study concentrates on the second phase of the development. In 1994, 1429 dwellings were realised in Rijkerswoerd phase two.

The whole development was carried out under direction of the municipality of Arnhem. Besides the aspects laid down in the land use plan, the

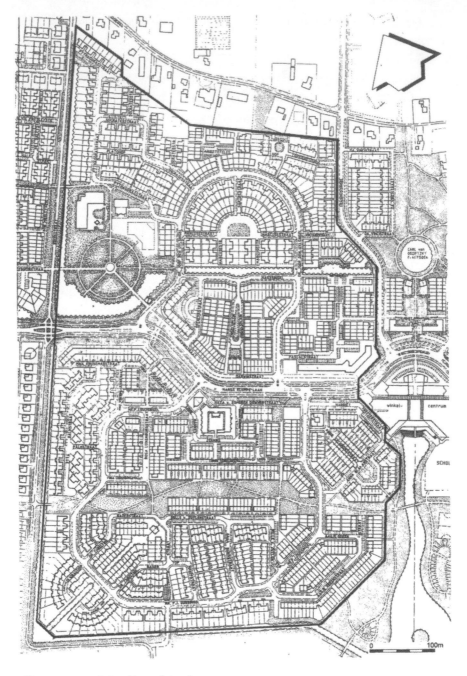

Source: municipality of Arnhem

Figure 2.2   Map of Arnhem Rijkerswoerd phase 2

municipality also told the house builders what type of houses, and how many of each to build. To this end, the second phase was divided into seven smaller sections of around 200 dwellings. The dwellings in each of these sections have been developed by a different private house builder or housing corporation. To realise the objective of 50% of the houses in the market sector and 50% in the social sector, the sections were developed in turn by a private developer (for market sector dwellings) and a housing corporation (for the social sector dwellings). For each section, the land use plan was further detailed by the developer. The municipality used the open spaces (roads, parks, green space, water) to create coherence in the whole area. It laid down its ideas for this coherence in the global land use plan, and has worked this out further in design guidance. The municipality imposed additional conditions on the developers, e.g. as to use of (environmentally friendly) materials, choice of architects, or choice of construction companies, when selling the building plots.

Zwolle Oldenelerbroek

Table 2.2   Core data Zwolle Oldenelerbroek

Location	The Netherlands, Provincie Overijssel, near Zwolle	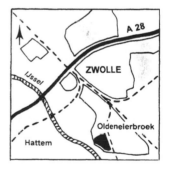
Size (ha)	39	
Number of dwellings	1068	
Housing density	27.4 dwelling/hectare	
Time period of development	1989-1998	
Initial ownership structure	Land in hand of a number of farmers, bought by the municipality just before development started	
Land developer	Municipality	
house builder	Private house builders, housing corporations	

51

Zwolle is a town in the east of the Netherlands, the capital of the province of Overijssel. It has about a 100,000 inhabitants, of which some 30,000 live in the area of Zwolle Zuid, a big urban extension of 11,000 houses to the south-west of Zwolle. This extension was finished in 1998. Oldenelerbroek – in the extreme south-west, the furthest away from the city centre – was the last housing area to be realised. The development of Zwolle Zuid started in 1978, on the basis of the Globaal bestemmingsplan Zwolle Zuid ('global' local land use plan) from 1974. In this plan, only the broad outline of Zwolle Zuid was given, for each part this had to be further specified in a more detailed local plan (called an Uitwerkings-bevoegdheid).

The decision to develop Zwolle Zuid was taken at the beginning of the 1970s. At the time, Zwolle was designated as a *groeistad* (growth city) in the national spatial planning report, which meant that it was to accommodate part of the regional growth. The municipality of Zwolle had compared different directions of development and Zwolle Zuid was chosen as the best solution. There was one important negative point to the development on this location: between Zwolle Zuid and the existing part of Zwolle there is an important barrier, in the form of the railway link Zwolle-Deventer.

The whole area of Zwolle Zuid was not developed at once. Since 1978 it has been developed in phases, beginning with the parts nearest to the existing built-up area. For each phase – usually between 1000 and 2000 houses – a separate land use plan was made, that detailed the initial 'global' local plan from 1978. In this global plan, only the main structure of the area was indicated. According to the wishes of the municipality and the market situation at that moment, the detailed plans were different. So the development of Oldenelerbroek started in 1989 with the specification of the global land use plan for that area. This designated the number and types of houses, housing densities, and the main infrastructure network. On the basis of this plan, the municipality developed a *beeldplan*, a vision of the future design of the area.

The layout of the area was specified in detail, but was not fixed once and for all in the *beeldplan*. Nevertheless, the only influence of the house builders on what has finally been built was the design of the houses. Because they had this freedom, they also had some influence on the

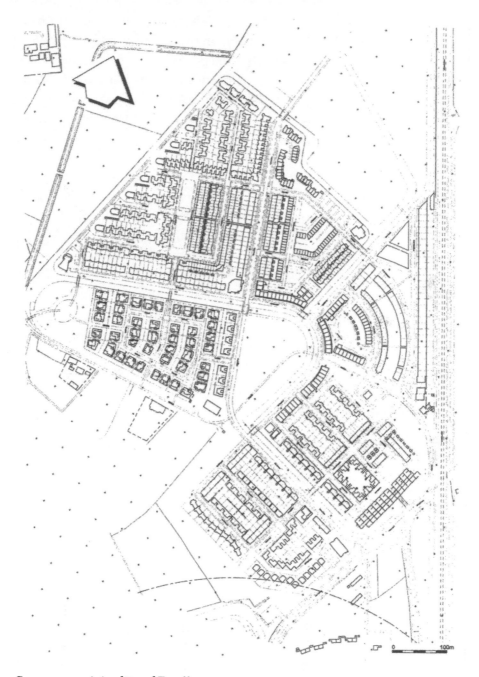

Source: municipality of Zwolle

Figure 2.3   Map of Zwolle Oldenelerbroek

53

parcellation. The municipality did not know exactly what type of houses the developers would choose to build, and therefore based the proposed parcellation on a standard house type and plot size. The house builders had the possibility to vary this a little, but they had to respect the road layout and the other principles laid down in the *beeldplan*. The further refinement of the *beeldplan* was done by the house builders together with the municipality in *projectteams*.

The whole area of 1100 houses was divided into several sections of some 200 houses each, that were assigned to different developers (either private house builders or housing corporations, depending on the type of houses). For each section, there was a project team that followed its realisation. In these teams the different departments of the municipality concerned with housing development were represented. Each of these had their specific wishes and demands. These were discussed with the wishes and demands of the developer, until an a form of the development acceptable to all parties was found.

During the planning phase, the municipality was also engaged in land acquisition. The land was not acquired a long time in advance, but just before it was going to be developed. The advantage of this was that interest charges were kept low, which meant that the selling price of the plots could be kept low. It was not necessary to use expropriation rights. Once the land was acquired, the municipality carried out the land development. Once serviced, the building plots were sold to the developers. At the start of the negotiations with a developer, the municipality signed a *project-overeenkomst* (project-agreement) with the developer. The developer then had to pay 10% of the price of the land. If the negotiations with the municipality in the project team did not work out the way the developers wanted, the latter could still withdraw, but then they would lose the money already invested.

Table 2.3   Core data Bishop's Cleeve

Location	The United Kingdom, Gloucestershire, near Cheltenham
Size (ha)	56
Number of dwellings	1750
Housing density	31.3 dwelling/hectare
Time period	1987-1998
Initial ownership structure	Partly agricultural landowners, partly two private housebuilding companies
Land developer	Mainly the two land owning private house building companies, some other house builders
House builder	Mainly the two land owning private house building companies, some other house builders

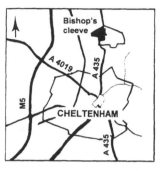

Bishop Cleeve is a village situated about three kilometres to the north of Cheltenham, a town with about a 100,000 inhabitants in the south-west of England. Before the housing development that this text is about started, there were between 1500 and 2000 dwellings in the village. Most of the residents of the newly developed housing in Bishop's Cleeve commute to Cheltenham. The local authority responsible for development control in Bishop's Cleeve is Tewkesbury Borough Council. This covers seven parishes of which Tewkesbury and Bishop's Cleeve are the biggest. At the beginning of the development process, parts of the land in the development area were already owned by two private house builders.

At the end of the 1960s, when growth expectations were high, much population pressure on houses was expected in the Severn Valley, the prosperous area in which Bishop's Cleeve and Cheltenham are situated.

Bishop's Cleeve was singled out by the central government as an area for growth. That was the start of the planning phase for a new housing scheme at Bishop's Cleeve, but it took a long time before this led to tangible results. Bishop's Cleeve was situated then on the main road between Cheltenham and Evesham. The function of that road has since largely been taken over by the M5 motorway between Bristol and Birmingham, but at the time, there was a lot of traffic passing through Bishop's Cleeve. The new plan was an opportunity to build a bypass around Bishop's Cleeve, to get the traffic out of the town centre.

In 1979, the County Council presented with its view of the future development of the area in the Draft Gloucestershire Structure Plan. In this plan, it proposed a development of 400 houses in Bishop's Cleeve and 600 to the south of Tewkesbury. There was strong objection to the latter, especially by the private house builders. That led to a reconsideration of the plan, and the County Council decided to shift the allocation of 600 houses from Tewkesbury to Bishop's Cleeve. That meant 1000 new houses in Bishop's Cleeve. In 1982, the structure plan was adopted in this form.

The fact that more houses were going to be developed in Bishop's Cleeve meant at the same time that house builders could provide the money for the by-pass. It created the need for an access road to this new housing, and it made the situation on the old road through Bishop's Cleeve yet more problematic.

In reaction to the County Council's plans, Tewkesbury Borough Council published the Cheltenham Environs Local Plan (CELP) in 1983, in which it made proposals for the accommodation of 1000 dwellings to the west of Bishop's Cleeve and for the routing of the bypass. In the CELP, the bypass formed a firm boundary to the housing development. In reaction, the house builders presented another routing for the bypass that allowed for housing construction on both sides. In a planning inquiry, the Borough Council lost the argument. In 1985, a planning application was submitted by the house builders, based on their plan for the development. In 1986 the revised CELP was adopted and following this, in 1987, an outline planning permission was granted for 1050 dwellings, subject to some land use conditions, a section 52 (now section 106) planning gain agreement with the Borough Council, and highway agreements with the County Council.

Source: Tewkesbury Borough Council

Figure 2.4   Map of Bishop's Cleeve (studied housing scheme within black line)

57

The planning gain agreements were a necessary complement to the CELP. Two house builders had acquired the land and they wanted to develop it for housing. The Borough Council had designated the land for housing, so in principle there were no objections against this. But the development of the area was subject to further negotiations, because it was said in the CELP that infrastructure facilities were to be 'funded with the aid of contributions from private developers'.

In 1987, the building started, based on the CELP and the agreements. During the development, the house builders built many more houses than the 1000 that were initially planned. The Borough Council felt it had no means to influence the housing densities. A problem that arose from the higher densities was that more open space than initially envisaged had to be provided. To realise this, in 1991 a new agreement was made between the house builders and the Borough Council, in which this issue was settled. In 1997, the development was in its final phase and there are likely to be around 1750 dwellings when the whole development is finished.

Table 2.4 Core data Cramlington North-East Sector

Location	The United Kingdom, Northumberland, near Newcastle	
Size (ha)	91	
Number of dwellings	2000	
Housing density	22 dwellings/hectare	
Time period	1980 – 1997	
Initial ownership structure	Land owned by two private house building companies	
Land developer	The two private house builders that were landowner, some other private house builders	
House builder	The two private house builders that were landowner, some other private house builders	

Cramlington is sometimes referred to as 'the first private house builder's New Town in the United Kingdom'. In 1997, it had around 30,000 inhabitants. It is situated at 15 kilometres to the north of Newcastle, a city with about 250,000 inhabitants. Newcastle is part of an urban conurbation with about 500,000 inhabitants. The proposal for a new town in Cramlington was first put forward in 1959 by Northumberland County Council. The development of the new town of Cramlington started at the beginning of the 1960s. In 1997, it was almost finished, and a total of 15,000 houses have been built. We focus on a part of this development, the North-East sector, realised between 1980 and 1997.

The County Council had political reasons for realising a large amount of housing here. One was to upgrade the area that was rather run down because of the decline in coal mining. Another was the location near the Newcastle conurbation. By developing this area, inhabitants from

Northumberland could benefit from the proximity of the city for employment. In 1962, a Comprehensive Development Area (C.D.A.) scheme was submitted to the Minister of Housing and Local Government, to be approved in 1963. This is the statutory local plan for the development that allocates the land uses in the area. In 1970 it was updated.

It is not clear who owned the land at the time the plans for Cramlington were made. The representative of the house builders says that the house builders owned the land before there were any plans for the development of the area. However, according to the representative of the Blyth Valley District Council, the house builders bought the land only after the plans for the development of the area were made. Apparently, the developers and the planners both had, more or less at the same time, the idea of building houses at Cramlington.

A rough phasing for the development of Cramlington was proposed in the original C.D.A. scheme. The proposed order of development of the residential areas was roughly first the South-East sector of the town, then the North-East sector, and finally the South-West sector. In the North-East sector, development started in 1980.

The last dwellings in this sector were completed in 1997. In that same year, the South-East sector had been entirely finished, the South-West sector was only partially developed.

The C.D.A. plan is very short. A big part of it is formed by the map, which shows the principal road network and the type of development which is to take place at each location. It sets only broad conditions. For that reason, the need was felt to make additional arrangements as to how the development was to take place, and what were the tasks of the different participants. That resulted in a private legal agreement. The signing of this agreement, usually referred to as the '1974 agreement', was an important event in the development of Cramlington. At the time, the notion of 'planning agreements' as presented in the 1991 Planning Act did not exist. Nevertheless, the parties managed to draw up an agreement that satisfied them for a long time, and that was revised only in 1994.

The agreement concerned three different points. The first one was the transfer of land by the developers to the local authority for the realisation of schools, roads, public open spaces, public services. The price that the County and the District Council had to pay for the land was also fixed in

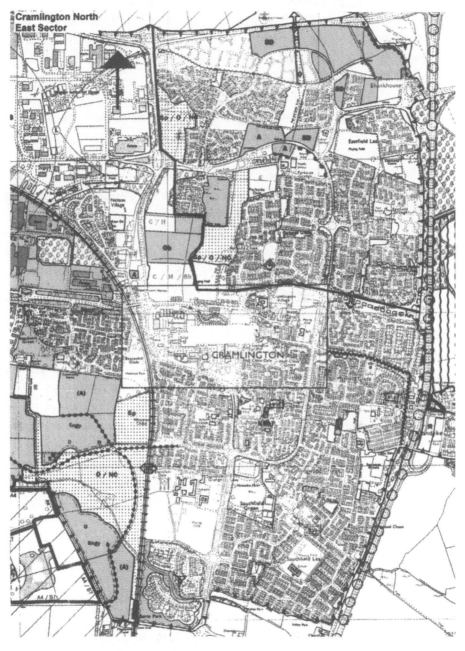

Source: Blyth Valley District Council

Figure 2.5 **Map of Cramlington (North-East Sector within black line)**

61

the agreement, at 11,400 Euro per hectare, which is little more than a symbolic price. Then there was a whole section about the drainage that had to be provided by the developers. The third point was the contribution to 'development works', the 'acreage contribution'. For every acre they developed, the house builders had to pay an amount to the planning authority. This money was used for 'community facilities' such as schools and public open space. With regard to the public open spaces, in the agreement, the house builders agreed to take care of the 'grading and seeding' of the green spaces.

Within the framework set by the C.D.A. plan and the 1974 agreement, the private house builders who owned the land steadily developed the town over the years. They did that in the usual way private house builders develop greenfield housing, in sections of two to two and a half hectare at a time. In 1982, a new document was issued by the 'Planning and development services committee' of Blyth Valley District Council, concerning the development of the North-East Sector of Cramlington. This document, called the 'Cramlington North-East sector housing strategy' was an addition to the original C.D.A. plan. In the first instance, this document concerned the phasing of the North-East sector of Cramlington. The second concern was to ensure a 'reasonable choice of attractively designed housing'. This is directly relevant for the residential environment. It is worked out in two items. One is that the District Council has tried to influence the housing densities in the area. The other is the introduction of 'special design sites'. The introduction of housing densities did not have much effect in practice. The District Council had no means to impose housing densities on the developers. With the introduction of special design sites, the District Council tried to create more variation in the housing development in Cramlington. The developers were not always very happy with these sites.

A last thing that needs to be mentioned is the division between the two house builders. From the very beginning these two parties divided their interests in half and half. Both of the parties owned half of the land and both of the parties developed half of the land. They had legally agreed this division, and had fixed it in a contract signed between them.

Table 2.5   Core data Bonn Ippendorf

Location	Germany, Land Nordrhein-Westfalen, near Bonn
Size (ha)	6.8
Number of dwellings	230
Housing density	33.8 dwellings/hectare
Time period	1985 – 1998, three years delay between 1986 and 1989
Initial ownership structure	over 60 original landowners
Land developer	municipality
House builder	original landowners, private house builders

Ippendorf was originally a village and has entirely been taken up in the agglomeration of Bonn (around 300,000 inhabitants). The residential development that this text focuses on is nearly seven hectares, and is divided into 115 building plots (which means there are around 230 dwellings, we return to that below). It is situated about eight kilometres to the south-west of the city centre of Bonn, and four kilometres to the west of the *Regierungsviertel*, the area of the city where the German Central Government had its offices until recently. After the events of 1989, Berlin became the new capital of a reunited Germany, and it was decided that the government offices should in time move from Bonn to Berlin. Until the time of the studied housing scheme in Bonn-Ippendorf, as regards the items that are important for this study – the demand for housing, the land market – the city did not seem to experience many changes as a result of this.

In the *Flächennützungsplan* (Structure Plan) of the city of Bonn of 1975, the area of Ippendorf was indicated as an area for future house building. This was the first step in the development of Ippendorf. However, the *Flächennützungsplan* did not say anything about the period in which the development had to be realised. It left this decision to the City Council, which made it in 1979. With that decision, the local planning procedure started, which led to the publishing of a local land use plan in 1985.

This local land use plan formed the basis for the start of an *amtliches Umlegungsverfahren* (formal urban land reparcellation, see appendix) in 1985. The decision to use the procedure of the *Umlegung* to prepare the land for development in Ippendorf was a logical one. The ownership structure in the area was very complicated, with a lot of very small plots and around eighty different landowners. This meant that the land had to be reparcelled before development was possible. It also meant that it was very difficult for one actor to purchase the whole area and carry out the land development as a temporary landowner.

In the procedure of the *Umlegung* the municipality drew up a new ownership structure that corresponded with the local land use plan, and that allowed land to be returned to the owners on the basis of what they brought in. These proposals had as an unforeseen result that the land use plan came again under discussion. The site where construction was to take place is on the bank of a small stream. The area nearest to the stream was never meant to be built upon, as it is protected under the *Bundes-naturschutzgesetz* (Federal nature conservation law). It was designated as woodland and farmland. But now the discussion started whether or not the whole plan should be abolished because of the proximity to the stream. The required revision of the land use plan took three years, a period during which the *Umlegung* could not be carried further because this needs a valid land use plan.

After this delay, the procedure of the *Umlegung* continued. In 1991, it resulted in a definitive parcellation, in which ownership structure, roads, play spaces, etc. were fixed. This *Umlegungsplan* consisted of a map and a *Verzeichnis*, a register of the new ownership structure. With the completion of the *Umlegungsplan*, the formal part of the development process ended, and the land development could start. In the first instance, this was financed out of the general budget of the municipality. Later, the costs were recouped from the landowners by means of the *Erschließungs-beiträge* (land development contribution). According to the federal building code, this contribution allowed the municipality to recoup 90% of the costs of putting in the primary services. The remaining 10% was paid by the municipality. Note that the municipality obtained the required land for the services free, through the *Umlegung*. Once the area had been serviced, the landowners started to build on their plots. They often did that themselves. Some owners sold their plots to others interested in building houses, but there are no big house builders that have bought large areas of land in Ippendorf. The biggest obtained four plots, on which it developed 16 dwellings. According to our informant, on average two dwellings per plot have been realised, hence a total number of 230 dwellings.

The local plan contains some guidelines as to how the houses had to be built, but the city of Bonn preferred not to prescribe too much, thus leaving the owners a substantial *Baufreiheit* (freedom to build). Indicated for each house (that is to say, for each plot) are whether the houses are detached or terraced, the number of floors each house can have (sometimes a minimum, always a maximum), the proportion plot size/built-up land, and the orientation of the house. On each plot a *Baufenster* (building window) is also indicated. This specifies – with some margin – on which part of the plot the house can be built.

By 1998, almost all the plots in Ippendorf have been built upon. However, some *Baulücken* (construction gaps) remain. The plots are in the hand of the private owners, and if these decide not to sell their plots nor to build on them themselves, the municipality cannot affect that decision. Moreover, in the local land use plan, a children's play area and a green space were envisaged. It was up to the municipality to realise these. But apparently the municipality did not have the money at the right time, or had other priorities that resulted in these facilities not being realised at the stage when they were envisaged.

Stuttgart – in the south-west of Germany – is with around 560,000 inhabitants one of the larger cities in the country. It is the major city of the regional conurbation of *Mittlerer Neckar*, with a total of around 2.3 million inhabitants. The economic base of this conurbation is one of the strongest in Germany, with lower than average unemployment rates. During the past few decades the favourable employment market has attracted inward migration from the *Land* Baden-Würtemberg, as well as from other parts of Germany. This also has to do with the residential and recreational value of Stuttgart, which is considered to be high. Consequently, pressure on the land market is high, resulting in high land prices.

Source: municipality of Bonn

Figure 2.6   Map of Bonn-Ippendorf

Table 2.6   Core data Stuttgart Hausen-Fasanengarten

Location	Germany, Land Baden-Würtemberg, near Stuttgart
Size (ha)	14.7
Number of dwellings	820
Housing density	55.8 dwellings/hectare
Time period	1993 – 1999
Initial ownership structure	Over 80 individual landowners
Land developer	Landowners, gathered in an *Erschließungsgemeinschaft* (land development company)
House builder	Landowners, private house builders, social house builders

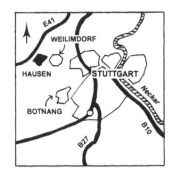

Hausen, some ten kilometres to the north-west of the city centre of Stuttgart, was a small settlement, with around 1000 inhabitants. The new scheme that we took as a case study adds around 820 houses, and 2,500 inhabitants to the built-up area of Hausen. Hausen is situated next to an attractive old estate, the *Fasanengarten*, which gives its name to the new development that is referred to as *Hausen-Fasanengarten 1 und 2*. The proximity of this estate – a big park – is good for neighbourhood recreation. For more extensive facilities, the district centre of *Weilimdorf* is two kilometres away, with regular bus links, and the *S-Bahn Haltestelle* (light rail station) to the city centre of Stuttgart is less than one kilometre away. The provision of a direct footpath from the housing scheme to the station is part of the plan for the development of the area.

The Municipal Council wanted to realise a large amount of social housing in Hausen-Fasanengarten. However, this was very difficult because of the high land prices in the area. Therefore, a means had to be found to moderate the price of at least part of the building land. In April 1993, a *Städtebauliche Entwicklungsmaßnahme* (urban development measure) was proposed by the planning department. However, before a municipality can decide to carry out a *städtebauliche Entwicklungsmaßnahme*, it has the statutory obligation to investigate whether the development might be carried out using another instrument. If possible, the use of a *städtebauliche Entwicklungsmaßnahme* should be avoided, because it is considered a very 'heavy' instrument, which among others gives the municipality substantial rights of expropriation for the designated area. This investigation led to the decision to develop the area by the use of an *Umlegung mit freiwillig vereinbarten Konditionen* (urban reparcellation on a voluntary basis). Thus, expropriation of the land – which would have taken place if a *städtebauliche Entwicklungs-maßnahme* had been used – could be avoided. Although the correct indication for this procedure is *Umlegung mit freiwilig verein-barten Konditionen*, for practical reasons it is often referred to, and will be referred to here, as *freiwillige Umlegung*. In fact, it was not the *Umlegung* that was 'voluntary', but the conditions under which it was carried out, the *Verteilungsmaßstab*. The building code offers the legal possibility to carry out an *Umlegung* using other conditions than the two specified in the code if all the landowners concerned agree upon these conditions (see appendix). That is what happened in Stuttgart.

In relation to this, it needs to be noted that the price of building land in Stuttgart is very high. In Hausen, the price of a serviced building plot was somewhere around 450 Euro per square metre. Since the municipality wanted to realise social housing in the area, it looked for a way to provide cheap land on which this would be possible. In the past, the municipality had developed a procedure of *freiwillige Umlegung*, at that time as a means to provide housing land quickly to respond to the high need for housing after the second world war. Now, it saw in this instrument a possibility to provide land at a reasonable price for social housing. Therefore, in June 1993, the Municipal Council asked the *Stadtplanungsamt* (department of city planning) to adapt and use again this old instrument. This procedure was chosen for the development of Hausen-Fasanengarten. It allowed the municipality to receive part of the value increase of the land caused by the development of the area in the form of cheap land and contributions towards the development of the area.

Because of the high value increase that resulted from the designation of the area for housing, the municipality considered it justified to make the landowners pay for the development of their land. To convince the landowners to do so, the municipality used the argument that if the landowners did not cooperate, the area would not be designated for housing, and hence there would not be any value increase. To get this argument through to the landowners, the municipality appointed an impartial intermediary body with no direct stake in the process. Basically, the landowners considered it reasonable to contribute to the development. However, they did not always agree as to the details of how much they had to pay, and for which specific services. Therefore, it was important that they had confidence that the municipality did not ask too much from them. For that confidence, the intermediary body was of great importance.

To arrange for the land development being carried out and paid for by the landowners, an *Erschließungsgesellschaft* (land development society) was set up with all the landowners, and this acted as a principal for the development works. This private land development in Hausen-Fasanen-garten is very particular. Usually in Germany, the municipality pays for the land development, the costs of which it later reclaims from the landowners by means of the *Erschließungsbeiträge*.

The intermediary body functioned as representative of the *Erschließungsgesellschaft*. Once the land development was finished, the works were adopted by the municipality. The latter had set standards up to which the area had to be serviced, and had checked whether these were adhered to, both during the works and once they were finished. When it found that the standards had been adhered to, the whole infrastructure in the area was adopted. This was the end of the activities of the intermediary body in the development process. The *Erschließungsgesellschaft* itself also became redundant and was dissolved. The landowners now had serviced building plots and it was up to them to build houses on those plots, within the prescriptions given by the municipality in the land use plan. In 1998, the housing construction was not yet entirely finished, however, the outline of the area was clearly visible.

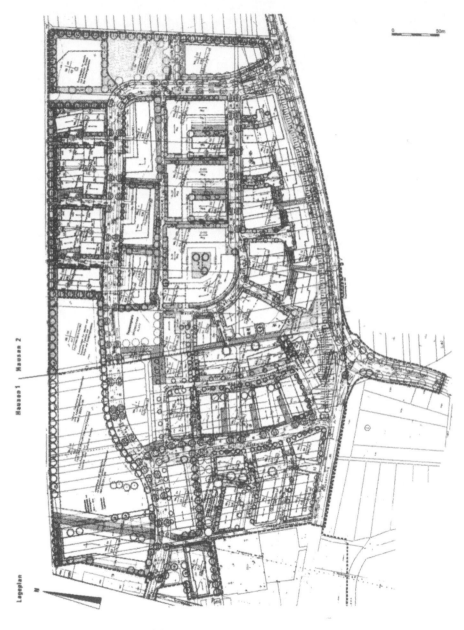

Source: municipality of Stuttgart

Figure 2.7   Map of Stuttgart Hausen-Fasanengarten

70

Table 2.7   Core data Bois-Guillaume Portes de la Forêt

Location	France, Région de Normandie, near Rouen	
Size (ha)	27	
Number of dwellings	410	
Housing density	15.2 dwelling/hectare	
Time period	1990-1998	
Initial ownership structure	Land owned by the municipality, acquired before from agricultural landowners	
Land developer	Private land development company	
House builder	Private house builders, social house builders	

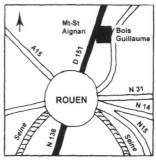

The operation 'Portes de la Forêt' in Bois-Guillaume is a residential development of 410 dwellings, on an area of 27 hectares. After these dwellings have been built, the operation enters into the next phase in which another seven hectares are to be developed with housing. This text describes the realisation of the first phase of 410 dwellings, which started in 1990 and was nearly finished by the end of 1998. The area also includes some commercial services, a generously dimensioned green area, and public facilities, such as a primary school and a nursery. It is situated in the municipality of Bois-Guillaume, a mainly residential community with a little over 10,000 inhabitants to the north of the city of Rouen, in the north-west of France. The city centre of Rouen is about four kilometres from the 'Portes de la Forêt'. Rouen is a medium sized city with about 150,000 inhabitants, on the banks of the river Seine.

As a result of an active policy of land acquisition in the period before the development of the Portes de la Forêt started, the municipality owned all the land in the area. This policy of land acquisition was supported by the procedure of the *Zone d'Aménagement Différé* (ZAD – differed development area). This procedure allows the municipality to declare the land as *utilité publique* (public utility), which gave it the right of pre-emption and if necessary of expropriation. It did not use these rights, but they probably worked as an incentive to landowners to sell their land.

The basis for the development of the 'Portes de la Forêt' was presented by the municipality of Bois-Guillaume in 1989 in the *Grandes options directrices d'aménagemen*' (general guiding principles for development). At this stage, the development of the area was proposed as a logical extension of the existing built-up area, to capture the expected residential growth. The form that the housing scheme should take was not dealt with. The municipality did not want to develop the area by itself. To choose the company to which the development was going to be delegated, the municipality launched a competition. It fixed the broad outlines of the development – the main transport axes, the boundaries, an approximate housing programme – and asked two private developers to propose how they would develop the area, bearing in mind these outlines. These private developers also had to take into account the financial balance of the operation. On these bases they elaborated a plan for the area. Out of the two plans, the municipality chose one and the developer who had drawn it up became responsible for the development of the area. This developer was what is called in French an *aménageur-lotisseur*. The English translation that comes closest to this is a land developer or subdivider. His aim is not to build and sell houses, but to develop and sell building plots. The house building is left to the purchasers of the plots.

As a legal framework for the operation, the procedure of the *Zone d'Aménagement Concerté* or ZAC (comprehensive development area, see appendix) was used. This procedure offers the possibility for negotiations between the municipality and private partners in the development about what exactly is going to be realised, and what financial contributions the developer should make towards public services and other aspects of the residential environment.

In 1991, there were objections against the plan, raised by a number of municipal councillors. They objected because the development plan was not in line with the *Schéma Directeur d'Aménagement et d'Urbanisme* (SDAU – general urban development scheme) for the agglomeration of Rouen. The SDAU proposed a northern bypass for the city that was not incorporated in the development plan for the Portes de la Forêt. In the first instance, the objections were judged legitimate by the *tribunal*

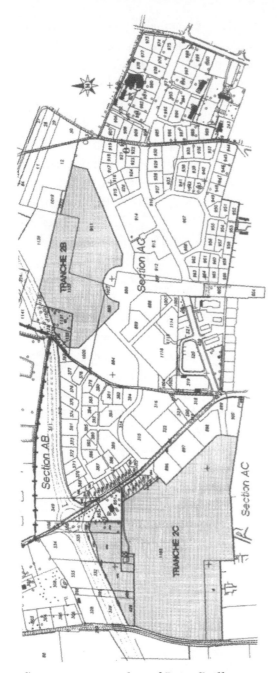

Source: municipality of Bois-Guillaume

Figure 2.8   Map of Bois-Guillaume Portes de la Forêt

73

administratif (administrative court). The municipality and the developer did not agree with this decision and went to appeal to the *Conseil d'État* (state council), which is the next higher jurisdiction. The latter decided that the SDAU on which the objections were based – and that dated from 1974 – did not correspond anymore to the situation at that time (1991). Therefore, the judge allowed the municipality to proceed on with the first part of the development. In the meantime the SDAU was to be revised and after that the second part of the development could be carried out. This led to a delay of about six months. After the *tribunal administratif* had judged the development not to be in accordance with the SDAU, the works could not continue, the development plan had lost its status. The developer and the municipality had to realise a new plan. In 1992 this was ready and approved by the *conseil d'état*, so the work could continue. The land development started in 1993, and in 1998 the development of the housing scheme was nearly finished.

Once the land developer had carried out the land development and had serviced the area, he sold building plots. All the areas that were not sold as building plots were transferred to the municipality once the land development was finished. In the agreement signed between the land developer and the municipality, it was agreed that the land developer had to maintain the green spaces for three years, before transferring them to the municipality. Besides giving this land to the municipality, the developer had to make more contributions to the municipality, in exchange for the right to develop the area. This was arranged in an agreement between the municipality and the developer. Thus, the developer contributed to a school and a nursery on the site, and to the extension of a secondary school, in proportion to the number of new pupils the area was expected to accomodate. And the developer was responsible for providing the area with public spaces, such as a central square, play spaces and green spaces, and for equipping them.

Table 2.8   Core data Rennes la Poterie

Location	France, Région de Bretagne, near Rennes
Size (ha)	110
Number of dwellings	2360
Housing density	21,5 dwellings/hectare
Time period	1980-1999
Initial ownership structure	Land in hands of the municipality, sold to the SEMAEB for development
Land developer	*Société d'Economie Mixte* (Public-private partnership)
House builder	Private house builders, *Offices de HLM* (housing corporations)

The development of the area of the Poterie was a logical next step in the overall development of the city of Rennes. In 1975, the municipality had finished the development of what is called the *ZUP Sud* (for: *Zone à Urbaniser en Priorité* – priority zone for urbanisation, south). This was a very large urban extension of around 20,000 dwellings, mainly high rise, as was usual for this kind of urban extensions at the time. The area of the Poterie was reserved for future extensions. Before the development started, the land was in agricultural use, but in the hands of the municipality. Since the beginning of the 1950s, the city of Rennes has followed a policy of land banking, buying in advance the land necessary for the urban development. To support this policy, the municipality used the procedure of the *Zone d'Aménagement Différé (ZAD)* (differed development area). This gives the municipality a right of pre-emption. For the development of la Poterie, the municipality sold the land to the *Société d'Economie Mixte de l'Aménagement et de l'Equipement de la Bretagne (SEMAEB)*. This is a public-private body to which the municipality of Rennes delegated the development of the area. The municipality and the SEMAEB worked in close cooperation, within the framework of the ZAC.

What was to be developed in the ZAC was fixed in a *Plan d'Aménagement de Zone* or *PAZ* (development plan for the area). The ingredients for this plan (the programme for housing, public facilities, etc.) were laid down by the municipality. The SEMAEB made the calculations to find out to what extent the wishes of the municipality could be realised taking all the financial aspects of the development into account. Once the PAZ was fixed, it was the task of the SEMAEB to realise it.

Due to its size, the development of the whole area took a long time. In 1998, the moment when the analysis upon which this text is based was carried out, the housing scheme was not yet entirely finished. Development of the seventh and last housing phase of the scheme, 510 houses on 6.5 hectares, started in the first half of 1998, and was due to be finished by the end of 1999. That would complete the development of the area.

The SEMAEB provided open spaces, roads, etc. Once the roads were finished, they were transferred to the municipality who was then responsible for the maintenance. Green and other public spaces were usually kept for one year in maintenance by the SEMAEB, as a sort of 'guarantee period'. Standards about the level of servicing of these areas were agreed between the SEMAEB and the municipality, and were checked by the municipality during the realisation of the scheme.

In the beginning of the development process, the number of houses brought onto the market each year had to be restricted. At that time, this was necessary to be able to sell them all. After this difficult start, which lasted some years, all the houses have been sold without much delay, and the scheme has brought jobs, facilities, and commercial activity to the area. Over the whole time span covered by the development, the initial objectives, both financial and in terms of the housing programme, have been realised. In spite of its slow start, the participants in the development process agree that on the whole, it has been a successful development.

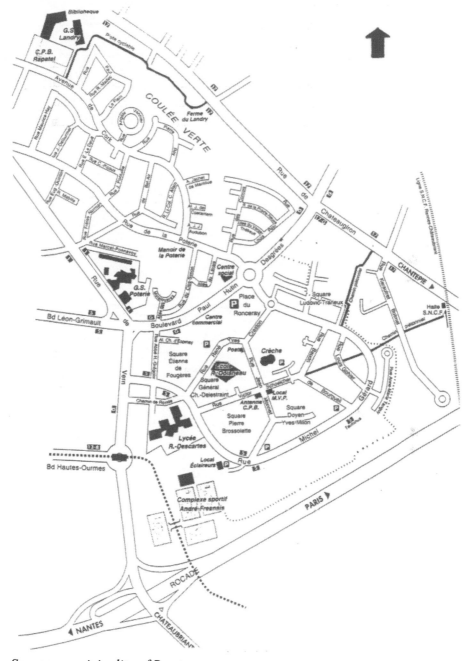

Source: municipality of Rennes

Figure 2.9   Map of Rennes La Poterie

3 The incidence of costs and revenues[1]

3.1 The residual theory of land prices

The basis for our financial analysis of the housing development process is the residual calculation of land prices. Ricardo introduced in 1812 in principles of political economy the principles behind this way of calculating the price of land. An often used quote of this author summarises the core of the residual land price theory: 'corn is not high because a rent is paid, but a rent is paid because corn is high; and it has justly been observed that no reduction would take place in the price of corn although landlords should forego the whole of their rent' (Ricardo, 1812: 38). In this chapter, it is first shown how a residual land price calculation works when applied to housing. Then we use it to bring to light the incidence and size of a financial margin in the different cases.

Land as a factor of production

In the residual land price theory, land is considered as a factor of production. The theory assumes that there is a market – and hence a price – for the product for which the land was a factor of production. The theory then states that the price of land is derived from the market price yielded by the product. That means that the price for housing land depends on the price of the houses that are built on it. The way this works is clearly summarised by Kruyt et al. According to them: 'A producer knows the

---

[1]   All figures from the case studies presented in this chapter are based on the financial accounts of the developers, where possible and/or applicable combined with the accounts of the local planning authority. All figures are in Euro. The exchanges rates that are used are: 1.- Dutch Guilder to 0.45 Euro; 1.- English Pound to 1.54 Euro; 1.- German Mark to 0.51 Euro; 1.- French Franc to 0.15 Euro.

price of a product X, and he can calculate the costs to produce X on a certain location if the necessary land were free of charge. (...) The difference between costs and price is called the residual value. If the producer could have the land at his disposal free of charge, this residual amount would be profits. If not, he will be prepared to pay an amount of money to obtain the land necessary to produce. This amount of money can mount up to the entire residual value' (1990: 15, translation RV).

The central assumptions of this theory have proven to be valuable and are still broadly used to determine the price of land prior to investment. However, throughout the years, Ricardo's initial ideas have not remained without criticism. It goes beyond the scope of this study to discuss the whole theorising about land prices. Others, like Mills (1972), Kruyt et al. (1990), or Nooteboom (1994) have done that elaborately. They argue that the residual theory should be seen as a behavioural approach, used by practitioners to determine the price of housing. Its reach is limited in two respects (see Nooteboom, 1994):

- it applies to a situation in which only one type of use of the land is permitted, i.e. in which alternative uses are prevented, for example by a planning system, or by the characteristics of the land;
- as a result, it applies to the short term. In the long term, the possibility of alternative uses cannot be excluded.

As a result, the residual land price theory cannot be used to analyse the changes in land prices in regions with many alternative possibilities, over a long time span. In addition to Ricardo's residual theory, economists have therefore sought for more sophisticated tools to analyse land prices. This resulted in what is called the neo-classical theory of land prices. Although this theory is not incorporated in our analytical framework, it gives useful background knowledge to understand the financial side of the case descriptions in this study.

The neo-classical theory analyses and explains land values using two central notions: transfer earnings and economic rent. The transfer earnings represent the price of a factor of production – for example: land – that is necessary to prevent it from transferring to other uses. In practice the price paid for the factor of production is often higher than the transfer earnings.

This surplus is called economic rent. The reason why a price higher than the transfer earnings is paid for factors of production is because they are scarce, relative to demand. Lipsey (1966: 350) gives an example to explain this: 'Movie stars (...) are in very short and pretty well fixed supply, and their possible earnings in other occupations are probably quite moderate; but, because there is a huge demand for their services, they may receive payments greatly in excess of what is needed to keep them from transferring to other occupations'. In much the same way, land for urban uses, for example for houses, is often in short and well fixed supply as a result of the planning system (see also Monk et al., 1991). Thus, an economic rent is often part of the price that is paid for land designated for urban uses.

The neo-classical theory has been criticised too, mainly because it searches for an explanation for land value only by looking at the demand side. Authors, such as Evans (1983) and Neutze (1987), have demonstrated the importance of the supply side in the determination of the price of housing. Both authors observe that the demand oriented explanations offered by Ricardian or neo-classical theory cannot explain certain features of the land market that can be observed. The main observation at the basis of the 'theory of supply' is that not all the land of which the price is above the agricultural value is used for urban uses, which should be the case according to the neo-classical theory. To explain this, Evans (1983: 128) presents a theory of land supply that takes into account '... the fact that occupiers of the land are unwilling to sell at the market price, but require some compensation for having to move their location. Because of people's 'roots' in an area, the compensation required may sometimes be quite large'. Neutze (1987: 387) gives three explanations for the same phenomenon: '(1) the non-financial direct utility which is obtained from occupation and ownership varies with its use and between owners, (2) land use decisions and land values depend on expectations about the future as well as on present conditions in the market, and (3) uncertainty about future conditions result in owner's expectations varying and the greater the uncertainty the greater the gains from deferring development.' But without disaffirming the importance of these explanations, Neutze states explicitly that they are additive to, rather than alternative for the demand oriented theory of land prices.

The residual land price calculation can be complemented by the stock adjustment model (Needham and Verhage, 1998). This model takes into account the neo-classical and supply side comments that have been made to the residual calculation of land prices. The stock adjustment model was applied to housing as long ago as 1960 (Muth, 1960) and has since then been applied to offices also (e.g. Rosen, 1983; Fischer, 1992). It is in general applicable when total supply of a product (regarded as a flow) is dominated by supply out of stock (i.e. second-hand goods being offered for sale). The goods transacted need not be uniform, differences in quality (e.g. between new and second-hand) leading to differences in price. But the general price level is determined by demand interacting with total supply, not with the supply of new goods.

If the price of new goods is above their production costs, suppliers of new goods will produce more. This will increase the total supply. If demand is constant, prices will fall. The size of that effect will be affected by the share of new supply in total supply: the smaller that share is, the smaller will be the effect on price of an increase in new supply. If the price of new goods is permanently and significantly higher than the costs of producing them, something is preventing stock adjustment to take place. That could be constraints on the capacity of the producers of new goods, keeping the flow of supply below the flow of demand, or a few producers restraining supply in their own interest. If prices are higher than costs for a long time, and markets are 'contestable' (Willig, 1987), then producers will want to expand their production capacity, and prices will eventually fall to the level of costs. If this does not happen, then there must be an external constraint.

When the stock adjustment model is applied to housing, certain properties of the second-hand housing market should be taken into consideration. The first is that some of the supply of second-hand housing is insensitive to prices: this is supply because of deaths, of two households amalgamating, or a household using less space, or moving out of the country. The second is that some of the supply is by existing owners 'trading up': they are sensitive to price differences rather than to absolute prices (for if the price of the desired house rises, then so does the price of the house to be sold to finance the purchase); but they do not like price falls (because that diminishes the value of their capital assets). If the price of housing is so high that developers build a lot, whereby prices start to fall, then existing owners thinking of trading up can delay or withdraw their plans, whereby the supply of housing falls, thus keeping prices up (see Janssen et al., 1994). The effect is not preventing the stock adjustment process, but that the adjustment takes place less smoothly.

Another particularity of housing is that it is not a uniform good, and there are some types of housing which it might not be possible to build, even though it would be profitable to do so. It is conceivable, for example, that some people prefer old housing to new: but old housing cannot be built. Another example is housing in a particular location, which is fully built-up (e.g. near to a park, or in an old and established suburb). Most types of housing which are in demand can, however, be produced in due time; and if the stock adjustment process can work itself out, production will continue until the price of those types of housing has fallen to its production costs.

Part of the stock adjustment model is the production cost of the goods. Applying this concept to housing is not straightforward. One reason is that a new dwelling requires infrastructure which it shares with other dwellings. So we need to consider the average cost of a dwelling which is provided as part of a large development. Another difficulty with the concept of the production cost of housing is the price to give to the land input. This should be the price necessary to transfer it from its existing use: this is the market value in that use less the costs of removing the existing development. Thus, the production cost of housing can be regarded as consisting of:

–   the transfer value of the land used;
–   the cost of servicing the land in order to transform if from its old use to building plots for the new housing;
–   the costs of building the new housing;
–   the profits which the suppliers of land and of housing need to induce them to supply.

Here lies a link between the stock adjustment model and our financial analysis based on the residual land price theory. The stock adjustment process takes place if the price of housing is above its production costs.

According to the residual theory, the 'extra price' that results from an external constraint on the supply of housing will be expressed in a higher residual value of building land. The description of the composition of the price of housing below shows how a high residual value can create a financial margin in the housing development process. Thus, through placing an external constraint on the supply of housing, spatial policy can influence the financial margin in the housing development process, hence the possibilities for expenditure on the residential environment (see also Evans, 1991).

Because of the conditions that are required for spatial policy to influence the stock adjustment process, this has more to do with national than with local policy for urban development. The policy needs to be applied over a wide area, and it needs to be applied consistently and steadily over many years. For that reason, the stock adjustment model is not part of the analytical framework for the empirical work, which focuses on the local level. Or, in other words, constraints on expanding supply are taken as given.

The composition of the price of housing

Although the theorising about land prices is both interesting and of great consequence, it is not our primary concern here. For the objective of this study, the analysis of particular housing development processes, the above-mentioned conditions for the use of the residual theory are met: only one type of use of the land is permitted, and our analysis applies to the short term. Therefore, the residual theory is used here, with the aim to explain the incidence and size of a possible financial margin in housing development processes. The broader explanation as to how this particular price came about when placed in its larger (socio-economic) context is not part of this study.

The scheme in table 3.1, an adaptation of similar schemes by De Kam (1996), and De Greef (1997), shows the composition of the price of housing, following a residual reasoning. This means that the price of the houses is taken as a starting point. The values of serviced building land and of raw building land are deduced from this price by subtracting the costs that have been made to realise the housing. The scheme is built up by combining the residual land price theory with the observation that a number of markets are involved during the development process of housing (e.g. De Greef, 1997; Priemus, 1998). The different markets distinguished here are the market for unserviced building land, for serviced building plots, for new housing, and for second-hand housing. A financial margin can appear in various places, namely on each of the different markets that are distinguished.

The scheme can also be seen as a representation of the development process. The activities in the development process that are distinguished in the analytical framework in section 4.1 (table 4.2) can equally be distinguished here. The first two activities in the framework in table 4.2, previous land use and land assembly, are combined here and terminate with the sale of the unserviced building land. Then starts the second activity, the land development, at the end of which serviced building plots are sold. This usually also marks the start of the next activity, that of housing construction. When the construction is finished, the houses are sold or let. Then starts the last activity, the ownership or use of the houses. Thus, this scheme corresponds with the framework for an institutional analysis of the housing development process presented in the next chapter.

It is possible that all of the four types of financial margin distinguished in the scheme appear in a certain development process, and that each of them is received by another actor. However, the financial margin in the lower part of the scheme is crucial for this. It can take up all the margin that might appear in the process. A simplified example (because it takes no account of uncertainties and changes during the process) can illustrate this. Suppose that at the start of a development process, the land is owned by a speculative landowner. This landowner will use a residual calculation to determine the maximum price he can ask for the land. In other words, he calculates the maximum financial margin on the market for unserviced building land. That situation is reached when higher in the scheme, the revenues are no higher than sufficient to cover all the costs (including the regular profit margin). If these are not covered, the land will not be developed, so the speculative landowner will not be able to sell the land. In the situation of a maximum margin on the market for unserviced building land, this margin eats up all the financial margin on

Table 3.1   The composition of the price of housing

value of a house
-/- financial margin on the market for second-hand housing[1]
revenues from sale of house
-/- construction costs[2]
-/- costs of finance and transaction
-/- regular profits[3]
-/- financial margin on the market for new housing[4]
revenues from sale of building plots
+/+ subsidy[5]
-/- costs of land development[6]
-/- costs of finance and transaction
-/- interest costs[7]
-/- regular profits[8]
-/- financial margin on the market for serviced building land[9]
revenues from sale of building land
-/- costs of finance and transaction
-/- interest costs[10]
-/- regular profits[11]
-/- financial margin on the market for unserviced building land[12]
value of land in existing use

Notes
1. The scheme starts with the value of the house on the market. If this value is higher than the price that is paid for it when it is first sold, the buyers of the house receive a potential financial margin. This occurs for example when a local authority fixes a maximum price for which the houses can be sold. This financial margin only appears when the house is actually resold for a higher price, in the market for second-hand housing. Since this study is concerned with new housing development, the financial margin on the market for second-hand housing does not play a role in the analysis.
2. A broad notion of construction costs is used here: all costs that are incurred during the construction of a house (also all fees, interest charges, etc.) are included.
3. The actor who is responsible for the construction of the house wants to have a compensation for his efforts in the form of profits, that is, he will require a certain level of returns on invested capital. The profits that are required depend upon the (economic) risks the house builder takes in realising the housing. If

86

the house builders cannot realise the required regular profits, they will not build the houses. The regular profits are therefore considered as an item on the costs' side in the residual calculation.

4. When all the construction costs are subtracted from the revenues of selling the house (the house price), and the difference is bigger than the price that has been paid for the building plots, then there is a financial margin on the market for new housing. This margin is received by the actor who builds and sells the house.

5. On several occasions during the housing development process, subsidies can be used. These are added to the revenues, which means that they allow raising the costs, or increasing a possible financial margin. Subsidy is mentioned here in the scheme because it occurs in some of the case studies (see section 3.4).

6. Analogous to the construction costs, a broad notion of development costs is used here.

7. Interest costs play a role whenever the actor who is responsible for the land transformation has the ownership of the land for a certain period, before the actual land development starts. During that period, money is invested in the land that does not yet yield any revenues. In other words, the actor pays for the interest costs on the money.

8. The same argument as in point 3, applies here, but with regard to the process of land development.

9. When the price of acquisition of raw building land and all costs for land development are subtracted from the revenues out of the disposal of the building plots, and the difference is positive, then there is a financial margin on the market for serviced building land. This is in the first instance received by the actor who is responsible for the land development, i.e. the (temporary) owner of the land during land development.

10. It sometimes occurs that an actor temporarily owns the land destined for housing development, but that the land is sold to another actor who carries out the land development. In that case (if possible), interest costs are charged here.

11. To sell his land as raw building land, a landowner might want certain profits above the value in the existing use, to compensate for risks or uncertainties this transaction implies.

12. Whenever first landowners receive more for their land than the value in the existing use plus the above-mentioned profit margin, then there is a financial margin on the market for raw building land.

the other markets. In the same way, the actor who sells the serviced building plots can maximise his financial margin, hence taking up the space for a financial margin on the markets higher up in the scheme.

For reasons that are elaborated in section 1.6, this study focuses on the process of land conversion, that is, on the land assembly and land development. For the residual land price calculation, described above, this means that our financial analysis is restricted to the lower half of the scheme. The financial margins on the market for unserviced and serviced land are what concern us here. To show the consequences of the residual calculation described above, the financial analysis of the eight case studies is presented and explained in the next section.

3.2   Case study files: financial analysis

Arnhem Rijkerswoerd phase two

When the municipality acquired the land in Rijkerswoerd, at the beginning of the seventies, it was used as farmland with an agricultural value somewhere between two and three Euro per square metre. The municipality bought it directly from the original landowners for the agricultural value. The municipality financed this with a loan. Therefore, the longer it took to realise the development, the more the interest costs rose.

The municipality temporarily owned the land and carried out the land development. It then sold the serviced plots to the housing corporations or to the private developers. Because the municipality acquired the land a long time before the actual development took place without paying any 'hope value', in this case there was no financial margin for the first landowners.

The interest costs, together with the costs of the land development, formed a large component of the price of serviced land. This price had to cover all the costs of the development. But of course it was partly dictated by the market: if the municipality had set the price too high, nobody would have acquired the land because it would not have been economically viable to build houses on it. This was not the case for the plots for social housing.

Table 3.2   Financial analysis of Arnhem Rijkerswoerd phase two

1429 dwellings in a plan area of 40 hectares 55% of the area sold off as serviced building plots (22 hectares)		
Agricultural value of the land (1.95 / m^2)	780,000	
Costs of land acquisition by municipality (1.95 / m^2)	780,000	
Financial margin (raw building land) 780,000 - 780,000		**0**
Costs of land development (33.90 / m^2)	13,560,000	
Preparation and management (3.80 / m^2)	1,520,000	
Interest costs (24.30 / m^2)	24,800,000	
Income from disposal of building plots (at an average disposal price of 63.60/m^2 serviced plot, i.e. 220,000 m^2)	13,995,000	
Income from Locatiesubsidie (central state subsidy)	9,990,000	
Total income from disposal of plots and subsidy	23,985,000	
Financial margin (serviced building land) 23,985,000 - 24,800,000 (possible regular profits for the municipality have not been subtracted)		**-/- 815,000**

These plot prices were determined by central government and could not be changed. In Rijkerswoerd the norm-price for social housing plots was not high enough to cover the costs of the development. The plots in the private sector therefore had to yield enough money to compensate for the loss on the social sector plots.

This contribution towards the social sector housing can be seen as part of the costs for land development. In addition, these consisted of all the costs for the primary and secondary services. The income from selling the building plots was not enough to cover all these costs. A *locatiesubsidie* (central government subsidy) was required to bridge the gap. In a way, this also means that there was no financial margin in the development of Rijkerswoerd. When the income generated by the land development appeared to be insufficient to cover all the costs, central government helped with a *locatiesubsidie*. This also implied that central government fixed some standards for the residential quality.

The house builders made profits only on the sale of the houses. No part of this profit margin was redirected into the residential environment. It needed to be sufficiently high to induce the developers to produce houses, but that has not been a problem in the case of Rijkerswoerd. The housing corporations – like the municipality – aimed only at covering their costs. Their construction activities can therefore be considered – like the house builders' activities but for different reasons – as not yielding extra money to spend on the residential environment.

Zwolle Oldenelerbroek

At the time when the land was acquired by the municipality, its agricultural value was about 2.70 Euro per square metre. However, the land had already been designated for housing development and had therefore gained a certain 'hope value', over and above its agricultural value. This resulted in a price of acquisition for the municipality of just over 10 Euro per square metre. After having acquired the land, the municipality developed it. The costs of the land development – putting in the services, connecting the existing infrastructure, etc. – mounted up to 18.40 Euro per square metre. When the costs of the management and preparation of the land development and the costs of site clearance are added, total costs of land development were close to 25 Euro per square metre.

In addition to the costs for land development, the municipality had interest costs. These were estimated at 15% of the total costs of the development of this area, i.e. 5.54 Euro per square metre. The rather high level of the interest costs can be explained because in the beginning of the 1980s, the municipality had purchased a lot of land in the whole area of Zwolle Zuid, after which the demand for housing fell dramatically. This meant that the municipality could not develop the land, because there was no market for building land. As a result, during this period the municipality received no income from the sale of building plots, but it did have to pay the interest costs on the land. Since the municipality works with one budget for the whole area of Zwolle Zuid, this increased the part that the interest costs played in the development of Oldenelerbroek.

The average price at which the municipality sold the building plots to the house builders was 86.72 Euro per square metre. In practice, the municipality did not use this price, it used plot prices. For social sector plots these were lower than 86.72 Euro per square metre, for private sector plots they were higher. For our purpose, the plot prices are converted to

Figure 3.1   Green space with children's play area in Zwolle Oldenelerbroek

prices per square metre. The revenues from the land disposal were used to cover costs of land acquisition and land development. A contribution towards secondary services is included and the revenues also had to cover the interest costs, mentioned above. What remains when all these costs are subtracted is the financial margin on the market for serviced building land. In the case of Zwolle Oldenelerbroek, a financial margin appears both on the market for unserviced building land, and on the market for serviced building land. However, regular profits for the first landowners and for the municipality have not been taken account of. It is reasonable to accredit the first landowners a certain profit margin. They need to be compensated for having to move somewhere else, to start a new business, or if that is not an option, to have a price that compensates for their lack of income.

Table 3.3   Financial analysis of Zwolle Oldenelerbroek

1068 dwellings in a plan area of 39 hectares		
56.4% of the area sold as serviced building plots (22 hectares)		
Agricultural value of the land (2.70 / m²)	1,053,000	
Costs of land acquisition by municipality (10.04 / m²)	3,915,000	
Financial margin (raw building land) 3,915,000 – 1,053,000		**2,862,000**
Costs of site clearance (0.81 / m²)	315,900	
Costs of land development (18.50 / m²)	7,215,000	
Contribution towards secondary services (3.60 / m²)	1,404,000	
Preparation and management (5.56 / m²)	2,168,400	
Interest costs (5.54 / m²)	2,160,100	
Total costs of land acquisition and development (44.05 / m²)	17,178,400	
Income from disposal of building plots (at an average disposal price of 86.72/m² serviced plot, i.e. 220,000 m²)	19,079,000	
Financial margin (serviced building land) 19,079,000 – 17,178,400		**1,900,600**

The financial margin on the market for serviced building land was received by the municipality. It was not the primary aim of the municipality of Zwolle to make profits. In this case, the municipality used the margin on this housing scheme to cover costs that were incurred in adjacent housing schemes that are part of the same urban extension of Zwolle Zuid.

Bishop's Cleeve

It is unclear when the house builders actually started to acquire the land in the area of the housing scheme. The informants said that land acquisition started not later than at the very first stages of the development process, but had probably already started before there were official plans. Nevertheless, because of the perspectives for growth in the area, the land certainly had a hope value that is expressed in the price of acquisition. The agricultural value of the land at the time of acquisition was around 0.95 Euro per square metre. The informants confirmed that the price of acquisition of agricultural land by the developers was around 28.50 Euro per square metre.

The price of the serviced land is hard to deduce in the British situation because serviced plots are hardly ever put onto the market, so generally their price has to be estimated. However, in Bishop's Cleeve some serviced land was traded, so we know what price was asked for it, namely 152 Euro per square metre. We know that there are on average 30 dwellings per hectare. That means that the price of a serviced plot was near 50,700 Euro. An average house in Bishop's Cleeve was sold at around 92,500 Euro. Around ten percent of the difference of 41,800 Euro between land price and house price is profit margin on constructing the house. The costs of land development are estimated at 45 Euro per square metre.

The total financial margin realised in Bishop's Cleeve is the total income from selling all the houses minus the total costs, consisting of the acquisition of the land, the land development, the housing construction and the profit margin. The financial margin on the market for unserviced building land goes to the first landowners. The agricultural value of the land was around 0.95 Euro per square metre, whereas they were paid a price of approximately 28.50 Euro per square metre. This means that on every hectare they sold, they received a financial margin of 275,500 Euro. Even when a regular profit margin is subtracted, it remains clear that some of the farmers became a 'millionaire overnight'.

Still, the house builders did acquire quite early in the development process and the 'hope value' at that time was not as high as the final value of the land, once developed. So the house builders also received part of the financial margin. The house builders were responsible for the costs of the primary (or 'on-site') services. Besides this, the agreements with the local planning authorities required the house builders to pay for the bypass, provide cheap land for schools and parks, and to grade and seed the parks.

They provided the Borough Council with 13 hectares of land to be used for community facilities outside the area that is considered here. The price they had paid for the acquisition of this land – hence the value it represents – is not available. It seems reasonable to accredit it the same value as the land within the housing scheme of Bishop's Cleeve, that is 28.50 Euro per square metre. That means for the whole area an amount of 3.7 million Euro. Together with this land, the house builders paid 770,000 Euro to the Borough Council for the development of the area.

That means a total contribution of roughly 4.5 million Euro. Then, the house builders had to provide 2.2 hectares of land for free to the planning authority to build a school in the area, and 0.8 hectare for a playing field. The value of this land is considerably higher than that of the area discussed above, because this was serviced land, which would otherwise have been developed as housing land. Therefore it had the value of serviced building plots, which was 152 Euro per square metre. That means that the value of these 3 hectares was roughly 4.6 million Euro. Finally, the house builders had to contribute to the financing of the bypass. Information about their contribution towards the bypass is not available. It is therefore not possible to calculate exactly the financial margin that the house builders received. We can only conclude that they received 29.12 million Euro (or 52 Euro per square metre), minus their contribution to the bypass, and minus possible interest costs (see table 3.4).

Table 3.4   Financial analysis of Bishop's Cleeve

1700 dwellings in a plan area of 56 hectares (30 dwellings / hectare)		
Agricultural value of the land (0.95 / m^2)	532,000	
Costs of land acquisition by house builder (28.5/ m^2)	15,960,000	
Financial margin (raw building land) 15,960,000 – 532,000		**15,428,000**
Costs of land development (45.– / m^2)	25,200,000	
Costs of housing construction (110.– / m^2)	60,480,000	
Contribution towards secondary services (16.25 / m^2)	9,100,000	
Contribution to bypass	?	
Interest costs	?	
Regular profits (10% of income from sale of houses)	15,540,000	
Total costs land acquisition and development (153.– / m^2)	126,280,000	
Income from sale of houses (based on 30 dwellings per hectare at an average price of 92,500)	155,400,000	
Financial margin (serviced building land and housing combined): 155,400,000 - 126,280,000		**29,120,000**

It has not been possible to find out the exact price for which the private house builders acquired the land in Cramlington North-East Sector. But we have some indications. First, the land was not about to be developed at the time of the acquisition. Moreover, as farmland, the land was not of much value. There had been a lot of coal mining activity in the area. The land had not yet settled and there was a lot of subsidence. So the farmers were probably not too reluctant to sell. This would mean that the 'hope value' of the land was low. The informants confirm this when they speak of a price of acquisition of somewhere around 1.50 Euro per square metre.

The price of the serviced land is difficult to distinguish in the British situation, because serviced plots are hardly ever brought on to the market. Plots are only sold in combination with a house. For that reason, in this case we can only find the combined financial margin on the market for serviced land and the market for first-hand housing, by the following calculation: The average price at which a house with its plot was sold was about 90,000 Euro. At 22 houses per hectare (see table 3.5), this means a total revenue of 203.50 Euro per square metre. That is a huge increase compared to 1.50 Euro that was paid for the unserviced building land, at the beginning of the 1960s. The difference is partly spent on the land development and the housing construction, and on interest charges because the land had been bought a long time ago. But there must have been a fairly important financial margin. Most of this went to the house builders. However, where they paid more than the agricultural value, part of the development gain went to the first landowners.

The house builders were responsible for the primary services up to a level set by the local planning authorities. This is a normal part of the costs of development, which is paid 'as a matter of course' by the developers. The developers also bear a part of the costs of the secondary services. By means of the 'acreage contribution', and under agreements about a cheap transfer of land for schools and public spaces, local government recoups part of the development gains. The acreage contribution was rather low in Cramlington since at the time it was agreed, the whole concept of planning gain did not yet exist. Moreover, the sum that was agreed was not index-linked, so that it became more modest over the years.

It has not been possible to find the exact financial margin because of a lack of data. The figures that have been found are presented in table 3.5. We are sure that the financial margin was lower than the sum mentioned in this table, because the interest costs and the contribution to secondary services still need to be subtracted. The regular profits are probably considerable too, because of the speculative character of the land acquisition by the house builder.

Table 3.5   Financial analysis of Cramlington North-East Sector

2000 dwellings in a plan area of 91 hectares (22 dwellings / hectare)		
Agricultural value of the land (0.50 / m^2)	455,000	
Costs of land acquisition by house builder (1.50 / m^2)	1,380,000	
Financial margin (raw building land) 1,380,000 – 455,000		**925,000**
Costs of land development (50.– / m^2)	45,500,000	
Costs of housing development (100.- / m^2)	92,000,000	
Interest costs:	?	
Contribution towards secondary services	?	
Regular profits (10% of income from disposal of houses)	18,216,000	
Total costs of land acquisition and development	157,096,000	
Income from disposal of houses (based on 22 dwellings per hectare at an average price of 90,000)	182,160,000	
Financial margin (serviced building land and housing combined) 182,160,000 - 157,096,000		**25,064,000**

Bonn Ippendorf

In the case of Bonn Ippendorf, the financial course of the development process depends more on regulation than in most of the other cases. It is therefore necessary to elaborate further on the regulation that is applied, i.e. that of the *amtliche Baulandumlegung*.

The agricultural value of the land within the municipal boundaries of Bonn was in 1998 around 10 Euro. However, the land in the area where the development took place was considered as *Bauerwartungsland* (expected development land). Because it was indicated in the municipal structure plan as land for future housing development, it already had a 'hope value'. Experience showed that the value of land with this quality in Bonn was about a third of the value of serviced building plots. For Ippendorf, that meant that the value of the land before the development process started was about 58.60 Euro.

Speaking of the price of acquisition of the land does not really comply with the procedure of the *Umlegung*, because the land for housing development is not acquired by anyone. The owners retain their land. At the start of the procedure, a value is accredited to this land that corresponds to the value of comparable land that is designated for housing, but not yet readjusted and serviced (this is called *Rohbauland*). In Bonn Ippendorf, this value was 122.40 Euro per square metre.

So the complete area of 6.8 hectare – the so called *Einwurfsmasse* – was accredited a value of 68,000 square metres times 122.40 Euro, which is 8.3 million Euro. The landowners now also knew the value of their plot, namely 122.40 Euro times the size of the plot in square metres. At the end of the procedure they received this value back, either in land or in money, according to their preferences.

From the *Einwurfsmasse*, the land required for roads and primary services was subtracted. In total, 1.4 hectare of land were required for this. The municipality already owned 0.2 hectare of existing infrastructure. That means that it needed another 1.2 hectare of land, that is 17.2% of the total *Einwurfsmasse*. The owners transferred this free of charge to the municipality. As a result, the plot they received back after the *Umlegung* was 17.2% smaller than the plot with which they went into the procedure. The total area which was returned to the owners as serviced building plots was 6.8 hectare minus 1.2 hectare (17.2%), i.e. 5.6 hectare. This land is called the *Zuteilungsmasse* (share out quantity). This *Zuteilungsmasse* is serviced land. The value of it is again determined on the basis of transactions of serviced building plots. For Ippendorf it was set at 173 Euro per square metre. That means that the total value of the serviced site was 56,000 square metres times 173 Euro, which is 9.7 million Euro.

The servicing of the land was done by the municipality. The costs for land development were recouped from the landowners, but this was not part of the procedure of the *Umlegung*. The costs were first paid by the municipality, out of the general budget. After the work was finished, this was recouped from the landowners in the form of the *Erschliessungsbeiträge* (contribution to land development). However, the rules allow only 90% of the total costs to be recouped. These costs for land development amounted to 2.9 million Euro (90% of which is 2.6 million or 38.20 Euro per square metre, charged to the landowners).

The above description of the financial flows in the development process of Ippendorf allows us to distinguish the financial margins. As mentioned above, the increase in value between *Bauerwartungsland* (expected building land) and *Rohbauland* (unserviced building land) goes to the landowners. This is the value increase due to the indication of the area for housing development in a local land use plan. The land had a market value of 58.60 Euro per square metre. At the start of the *Umlegung*, it was accredited a value of 122.40 Euro per square metre, as *Rohbauland*. This means the landowners received a financial margin of 63.80 Euro per square metre. Note that they still had to pay the *Erschließungsbeiträge*, which reduced this margin by 38.20 Euro per square metre.

The municipality also received part of the value increase of the land. First of all, the municipality received the above mentioned 11,560 square metres (17.2% of the plan area) of *Verkehrsflächen*, land necessary for the servicing of the area. After this has been subtracted, a total area of 56,440 square metres of building plots remained. This had a value of 173 Euro per square metre, which means that the total value of the area after the *Umlegung* was 56,440 times 173, which is 9,764,120 Euro. The value of the total area before the procedure was 68, 000 square metres times 122. 40 Euro per square metre, which is 8,323,200 Euro. The difference between these two values (1,440,920 Euro), was paid by the landowners to the municipality. The municipality used this to finance the administration, the planning procedure, and the secondary servicing of the area. However, on the basis of a rough calculation, the informant at the municipality says that the city of Bonn 'gained' around 765,000 Euro on the development of Ippendorf. This can be considered as the financial margin received by the municipality.

After the procedure, landowners can receive a financial margin if they manage to sell their plots for more than the 173 Euro per square metre that was calculated in the *Umlegung*. They also receive a possible development gain on the construction of the houses. Whether this has been the case goes beyond the scope of this study.

Table 3.6   Financial analysis of Bonn Ippendorf

230 dwellings in a plan area of 6.8 hectares		
Agricultural value of the land (11.20 / m²)	761,600	
Market value of the land concerned (58.60 / m²)	3,985,000	
Value at which land is brought in (122.40 / m²)	8,323,200	
Financial margin (raw building land): 8,323,200 – 3,985,000		**4,338,200**
Out of this margin, 90% of the costs of land development still have to be paid, after which remains 4,338,200 – 2,560,000		**1,778,200**
Costs of land development (41.82 / m²) (90% of which is charged to the landowners)	2,844,000	
Extra contribution by landowners (Contribution of land converted to money)	1,440,920	
Income municipality 1,440,920 + 2,559,600 (all contributions by landowners)	4,000,520	
Financial margin (serviced building land): Municipality receives 4,000,520. This is used for financing land development and additional costs. Remaining financial margin:		**765,000**

Stuttgart Hausen-Fasanengarten

In order to understand the way the *freiwillige Umlegung* works financially, it is first necessary to understand the way in which the issue of land value is dealt with in Stuttgart. The land value is calculated in different stages. The first stage is that of *Agrarland* (agricultural land). This is land in agricultural use, without any expectation that it is going to be developed in the future. In Hausen-Fasanengarten, the value of the land in agricultural use (*Agrarland*) was 7.65 Euro per square metre.

When land is destined for development in a land use plan, it is called *Rohbauland* ('rough' or unserviced building land). This has a much higher value, because – in this case – it is going to be used for housing. In Hausen-Fasanengarten, the value of unserviced building land (*Rohbauland*) was 229.50 Euro per square metre. When the land was ready for construction, it had become *Bauland* (serviced building land), with a value of 450 Euro per square metre. When an *amtliche Umlegung* is used, the value increase from *Agrarland* to *Rohbauland* goes entirely to the landowner (see case study Bonn Ippendorf). The *freiwillige Umlegung* allows the municipality also to receive part of this value increase, and that is exactly the reason why the municipality of Stuttgart chose to use this procedure in Hausen-Fasanengarten. The following arrangement was made between the landowners and the municipality: each landowner had to give 30% of his plot free of charge to the municipality. This is the land required for roads, public spaces, etc. Then the landowners had to give 20% of their land for a reduced price of 102 Euro per square metre to the municipality (instead of the normal value of unserviced building land, which was 229.50 Euro). Moreover, the landowners were responsible for the servicing of the area, which amounted to 41.60 Euro per square metre.

The clearest way to illustrate how this worked out is by giving an example. Imagine a landowner who owned a plot of 2000 square metres of land in agricultural use in the area of Hausen-Fasanengarten. At the start of the housing development process, the land had a value of 7.65 Euro per square metre, so the value of the plot was 15,300 Euro.

The initial value of the plot of 2000 square metres was 15,3000 Euro (agricultural value). After the *Umlegung*, 30% (600 m^2) are subtracted from the plot for roads and public spaces. Besides, the landowners are obliged to sell 20% (400 m^2) for the price of 102 Euro per square metre to the municipality as *Sozialbeitrag*. After the *Umlegung*, they retain a plot of 1000 square metres. The value of this plot as serviced building land is 1000 times 450 Euro, which is 450,000 Euro. So although the plot is reduced in size, it increases in value by 450,000 minus 15,300, which is 434,700 Euro. Part of this value increase is attributable to the costs of land development, i.e. 2000 times 41.60, which is 83,200 Euro. Then the landowners had to contribute towards the financing of the *soziale Infrastruktur* (social infrastructure e.g. primary school, kindergarten, gymnasium), and they had to pay for compensation measures imposed by the *Bundesnatur-schutzgesetz* (federal nature protection act). The agreed amount of this contribution from the landowners was 48.50 Euro per square metre of building land. For the owner of this plot this means a contribution of 1400 (m^2 of building land) times 48.50, which is 67,900. Finally, the landowners had to pay the costs of administration and the fee of the GSL (the private body that managed the operation). Together, this adds up to 5.70 Euro per square metre of *Einwurfsfläche*, or 2000 times 5.70, which is 11,400 Euro in total.

Figure 3.2   **Part of Hausen-Fasanengarten nearest to the existing built-up area of Hausen**

When all the costs are subtracted from the value increase, the remaining financial margin for the owner of a plot of 2000 square metres is 434,700 minus 67,900 minus 11,400, which is 355,400. Note that this is the combined margin on the market for unserviced and for serviced building land, since there is no temporary landowner. Part of this margin can be considered as the regular profits.

The municipality receives a contribution towards secondary services, and an indemnification for its administration costs. Together this adds up to almost 70,000 Euro. Moreover, the municipality receives 600 m^2 of land free of charge, and it receives 400 m^2 of land at the reduced price of 102 Euro per square metre. In other words, for these 400 m^2 it captures a value increase of 450 minus 102, which is 348 Euro per square metre, or almost 140,000 Euro. It needs to be noted that officially, the municipality is not allowed to recoup planning gain. The part of the development gain it receives here is all related to costs the municipality incurs in the development process.

This example shows why both the municipality and the landowners are prepared to enter into negotiations within the *freiwillige Umlegung*. In table 3.7 this calculation is shown for the whole area of the development.

101

Table 3.7   Financial analysis Stuttgart Hausen-Fasanengarten

820 dwellings in a plan area of 14.7 hectares		
8.36 hectare building plots (56.9%)		
Agricultural value of the land (7.65 / m²)	1,124,550	
Value at which land is brought into the procedure (229.50 / m²)	33,736,500	
Financial margin (raw building land): 33,736,500 – 1,124,500		**32,611,950**
Costs of land development (41.60 / m²) (Charged to the landowners)	6,120,000	
Contribution by landowners in land for social uses (*Sozialbeitrag*, missed value increase for 20% of the land)	10,228,000	
Participation towards secondary services (*Soziale Infrastruktur + Ausgleichsmaßnahmen*)	3,932,000	
Total costs of landowners	20,280,000	
Costs of municipality for secondary services (50.80 / m²)	7,471,500	
Other costs of municipality (indemnification, management, land acquisition)	4,299,300	
Total costs municipality	11,770,800	
Estimate value of the site after development (8.36 hectare building plots at 450 / m²)	37,620,000	
Financial margin landowners (serviced building land): Value of site - costs made by landowners - value at which land is brought in = 37,620,000 - 20,280,000 – 33,736,500		-/- **16,396,500**

The negative financial margin on serviced building land is received by the same landowners as the financial margin on raw building land. When subtracted, the remaining financial margin for the first landowners is 32,611,950 - 16,396,500 = 16,215,450. Besides, the municipality had a negative net result of 7,838,800. That is the total of its contribution to the housing scheme.

Bois-Guillaume Portes de la Forêt

All the figures in this section are based on the provisional balance sheet of 1994. This differs from what has actually been realised. The income side of this balance sheet is based on the constrution and sale of 355 houses in different categories. In reality, 410 houses were built on the site. The presented figures represent the calculations that form the basis for the development. On the basis of these figures, the parties went into the process. It has not been possible to obtain the actual prices.

The agricultural value of the land in the area concerned was about 0.90 Euro per square metre. However, since the area was in a zone destined for future urbanisation, the land had gained some hope value. The municipality acquired the land at an average price of 6.60 Euro per square metre. The municipality was only an intermediary landowner. For the actual development, the land was transferred to the private land developer Foncier Conseil. The price this private developer paid consisted of the price for which the municipality had acquired the land, increased with the interest costs it had incurred in the period it owned the land. The price that resulted was 7.05 Euro per square metre.

The private developer was responsible for the land development. After development, the building plots were sold. The plot prices differed according to the type of dwelling. The differences in plot prices were due to the difference in size between the plots and the price discrimination between different types of plots. The average price paid for a plot (all types of housing together) is about 24,500 Euro. Besides the plots for houses, there were 2100 square metres of office space in the plan. This was disposed of at a price of 130 Euro per square metre, which means a total of 0.27 million Euro. That leads to a total revenue from the disposal of serviced land of roughly 8.7 million Euro (355 plots times 24,500 Euro per plot) plus 0.27 million Euro for offices, which is almost 9 million Euro. This is a little over 33 Euro per square metre of the total area (27 hectares).

The costs for the developer consisted of three parts. First, there were the costs for the primary servicing of the area. These amount to 3.4 million Euro. Then there are what is called in French the *participations*: the contributions of the developer towards the secondary servicing. The total amount of money the developer spent on the *participations* is 2.4 million Euro. The last part of the costs of land development consists of what is called the 'additional costs'. These include the financial costs, the costs of administration and management, a provision for unexpected events and a margin for the developer. This adds up to 1.25 million Euro. Added to this are the costs of land acquisition of 1.9 million Euro. Thus, the total costs of land development amount to 8.95 million Euro. That is 33.15 Euro per square metre for the whole area. Presented in this way, there is no financial margin (see table 3.8).

Table 3.8   Costs and income in Bois-Guillaume Portes de la Forêt

355 dwellings in a plan area of 27 hectares		
Agricultural value of the land (0.90 / m²)	243,000	
Costs of land acquisition by house builder (7.05 / m²)	1,903,500	
Financial margin (raw building land) 1,903,500 - 243,000 The interest costs incurred by the municipality for the carrying of the land still need to be subtracted		**1,660,500**
Costs of land development (12.61 / m²)	3,405,000	
Administration and management (1.19 / m²)	322,500	
Costs of finance (i.e. interest costs) (1.33 / m²)	360,000	
Contribution towards secondary services (8.83 / m²)	2,385,000	
Reserve for unexpected expenses (0.28 / m²)	75,000	
Regular profits (1.83 / m²)	495,000	
Total costs of land acquisition and development (33.15 / m²)	8,946,000	
Income from disposal of building plots	8,946,000	
Financial margin (serviced building land)		**n.a.**

As regards the financing of the residential environment, it is interesting to look at the *participations* in more detail. These *participations* have three main components. The biggest part are the financial contributions the developer makes to the construction of the school and the nursery on the site, to the extension of a secondary school outside the plan area and to the relocation of allotments that had to make way for the development. These financial contributions amount to 1.29 million Euro. Then there are the works that have not only been financed, but are also carried out by the developer. This concerns the construction of a market hall on the central square, the equipment of a play space and the construction of a flyover, over the bypass. In this category, the developer also calculated the loss of income caused by the disposal of land designated for social housing at a price below the market value. All together, this adds up to 0.83 million Euro. Finally, the developer contributed to the residential environment by offering the land for the school and for the play space free of charge. This cost 0.27 million Euro.

The municipality could have chosen to serve the area itself. To be able to pay for that, it would then have sold the land at a higher price to house builders. However, it was decided to have a developer carry out the works. The sale of the land for the price that was agreed can be seen as a financial contribution by the municipality towards the residential environment. But this contribution did not really cost it money, since it sold the land at the price of acquisition plus interest and other costs related to the temporary ownership of the land. Seen from a different angle, the *participations* by the developer replace the higher price which the municipality could have asked for the land.

Rennes La Poterie

The major part of the land for the development of the Poterie was deliberately bought a long time in advance of any development. Thus, the parcels that were bought were not yet connected to any urban networks (gas, infrastructure, sewerage, water). A *Zone d'Aménagement Différé* (exeptional development area) was installed in the area, which gave the municipality pre-emption rights. The price it had to pay when using this pre-emption right was the price of agricultural land. Even in cases of legal dispute, the land is judged as being agricultural land, hence the price of acquisition was the agricultural value of the land, around 1.– Euro per square metre.

At the start of the development, the land was transferred to the public-private developer, the *Société d'Economie Mixte pour l'Aménagement et l'Equipement de la Bretagne* or SEMAEB. The price that the SEMAEB paid for the land was 4.45 Euro per square metre. This price of acquisition with which the development process starts consists of the price of cquisition of agricultural land increased by the interest costs the municipality had incurred due to the temporary ownership of the land.

Figure 3.3   Detached dwellings in the older part of Rennes La Poterie

In a next phase, the SEMAEB carried out the land development. All the costs related to this had to be covered by the revenues from the sale of the building plots. The secondary services are partly financed by the municipality, partly by the SEMAEB in the form of *participations* (contributions) by the house builders.

There is one major exception to the general rule that everything within the limits of the plan is financed by the revenues from the disposal of the building plots. This is the main through road, the *Boulevard Paul Hutin Desgrées*. Its function is wider than the servicing of the Poterie. Therefore, it was taken out of the budget of the operation. There is, however, a contribution from the operation towards the road, because it does serve also the residents of the Poterie.

The contributions towards secondary services are not included in the same budget as the costs of land development. At the start of the operation, the level of servicing of the area was fixed in the *dossier de realisation* (realisation file). The costs of realising this fixed level of servicing had to be covered by the income from the sale of the building plots. Within the operation, cross-subsidising between profitable and non-profitable parts was possible. For example between the commercial centre, which was

106

profitable, to the social rental sector, which was non-profitable. A small contribution from the municipality was agreed to balance costs and income on the accounts.

Table 3.9   Financial analysis of Rennes la Poterie

2360 dwellings in a plan area of 135 hectares		
Agricultural value of the land (1.– / m^2)	1,350,000	
Costs of land acquisition by SEMAEB (4.45 / m^2)	5,995,200	
Financial margin (raw building land) 5,995,200 – 1,350,000		**4,645,200**
Site clearance (0.51 / m^2)	688,500	
Costs of land development (27.25 / m^2)	36,786,450	
Interest costs (3.60 / m^2)	4,860,000	
Development costs (i.e. regular profits) (2.29 / m^2)	3,088,350	
Total costs of land acquisition and development (38.09 / m^2)	51,418,500	
Income from disposal of building plots	43,813,050	
Other income (subsidies, contributions from developers and city, ...)	10,888,200	
Total income from disposal of plots and subsidy	54,710,250	
Financial margin (serviced building land) 54,710,250 – 51,418,500		**3,282,750**

In the course of the development process, the requirements as to the level of servicing were sometimes slightly changed by the municipality. In these cases, there was a renegotiation between the SEMAEB and the municipality. In cases where the costs of the new requirements were higher than what was initially agreed, the municipality sometimes contributed to the primary or on-site services.

The price at which the building plots are sold differs for the different categories of housing. The price is expressed in the price per square metre of living space (*surface habitable*). These prices are reviewed every year. The average disposal price of building plots is around 225 Euro per square metre, which is the *prix d'équilibre* (equilibrium price) of the operation.

3.3 The cost side

In this and the following sections (3.3 and 3.4), we consider in more detail the various components of the financial accounts of the housing development processes. First, we focus on the cost side. As can be seen in the case descriptions presented above, the costs in housing development processes are grouped under broad headings. In this section these broad items, which can generally be found in all financial accounts of housing development processes, are explained and analysed using data from the case studies.

Land acquisition

The activity of land acquisition does not occur in development processes where the first landowners retain the land during the development. But in all the other cases, the land is acquired by the developer prior to development. In our analysis, the costs of land development form an important variable. As explained before, a low price of land acquisition creates a financial margin in the process, which is then available for something other than for obtaining the land required for development. If the whole financial margin has to be used to acquire the land, this limits the possibilities for expenditure on the residential environment later in the process. Since all our cases concern greenfield development, at the start of the process the land is in agricultural use. But more than the agricultural value, it is the level of certainty whether the land is going to be developed that is crucial for the value of the land, hence for the price that is paid for the acquisition. Examples of cases with a particularly low, or with a very high hope value of the land illustrate this.

A very low price of land acquisition is found in the case Cramlington North-East Sector (table 3.5). The land in Cramlington was poor quality farmland. Nevertheless, the land was in agricultural use at the time it was purchased by the private house builders at the end of the fifties, beginning of the sixties. It has not been possible to find out the exact price of acquisition, due to the confidentiality of the information, but we know that the land was not about to be developed at the time of the acquisition. Moreover, as farmland, the land was not very valuable. So the farmers

108

probably were not too reluctant to sell. This would mean that the 'hope value' of the land, the value above the agricultural value due to the anticipation of a future rise of the land price resulting from the development of the area, was low. The informants confirm this.

Because the costs of acquisition were so low, they formed only a minor part of the total costs made during the land conversion, the costs for land development forming the major part. If we know the costs of acquisition and the costs of the development of the site and the price that is eventually paid for a plot with a house on it, we can estimate the financial margin. In the case of Cramlington, this was relatively high, because the costs of land acquisition were so low. Over the years, the land gained much more value, basically because it became designated as a new town. The difference is partly spent on interest charges, because the land had been bought a long time ago, but it is very probable that a large financial margin remained in Cramlington-North East Sector, mainly due to the low costs of land acquisition.

The low price for land acquisition that was paid in Rennes la Poterie in France (table 3.9) was not only due to the absence of development pressure at the time of acquisition. The major part of the land for the development of the Poterie was bought long before the development started. At that time, the land was in agricultural use. The municipality deliberately bought the land a long time in advance of any development, using pre-emption rights. The price it had to pay when using this pre-emption right was the price of agricultural land, although it was clear that in the long term the land would be used for urban development. But even in cases of legal dispute, the land was judged as being agricultural land because it was bought so early that it was not yet connected to any urban networks.

At the start of the development of the area, the land was transferred to the *Société d'Economie Mixte pour l'Aménagement et l'Equipement de la Bretagne* (SEMAEB). The municipality did not recoup any value increase of the land that might have occurred during the period it had the land in its possession. Because of the active policy of land acquisition and hence of the low price that the municipality had paid for the land, the SEMAEB could acquire the land for only 4.45 Euro per square metre.

The case of Stuttgart Hausen-Fasanengarten (table 3.7) shows the highest hope value for unserviced land of all the studied cases. Although there was no intermediary landowner who actually acquired the land, the problems that a high hope value can cause in the development process are well demonstrated in this case. The case is also a good example of a creative answer to these problems. We return to that in the next chapter.

The price of building land in Stuttgart is very high, both compared to other German cities, and as the part it represents of the total price of a house. In Hausen, the price of unserviced building land is near 250 Euro per square metre. The high price of the land, and hence the high price of building plots, poses problems for the realisation of social housing, for example. Since the municipality wanted to realise social housing in the area, it looked for a way to provide cheap land on which building social housing would be possible. Therefore, in June 1993, the municipal council asked the *Stadtplanungsamt* (department of city planning) to reinvigorate the old instrument of *freiwillige Umlegung* (which should actually be called: *Umlegung mit freiwillig vereinbarten Konditionen* (land readjustment on voluntarily agreed conditions)).

The *Umlegung mit freiwillig vereinbarten Konditionen* allowed the municipality to receive a substantial part of the value increase of the land due to the development of the area, in the form of cheap land and contributions towards the development of the area. The municipality received − above the 30% of the total area for infrastructure facilities, to which it is entitled in any *Umlegung* − another 20% of the total land area for a price of only 102 Euro per square metre. This land is used for social housing or community facilities (*Gemeinbedarf*). Above that, the municipality demands of the landowners that they pay for the costs of land development, which amount to 50 Euro per square metre.

The enormous value increase of the land when it changes from agricultural to building land creates a big financial margin on both the market for unserviced and for serviced building land. Rather than letting a particular actor capture this financial margin, in Stuttgart Hausen-Fasanengarten a way was found to deploy it in the development process. The value increase was so big that even after this, a considerable financial margin remained, which meant that the procedure of the *freiwillige Umlegung* offered a good solution in this case to cope with the high land prices. How and why this procedure worked is dealt with in chapter four.

110

The costs of land development are the costs that are made for providing the primary – or on-site – services on the development sites. The case of Rennes La Poterie (table 3.9) is used here to explain the role played by these costs. In this case, and in general, these costs had to be covered by the revenues from the sale of the building plots. This can be contrasted to the secondary services, that were in Rennes la Poterie partly financed by the municipality, partly by the developer (the *Société d'Economie Mixte pour l'Aménagement et l'Equipement de la Bretagne*) in the form of *participations* (contributions).

At the start of the operation, the level of servicing of the area was fixed in the *dossier de realisation* (realisation file). The costs of realising this fixed level of servicing had to be covered by the income from the sale of building plots, which is always the case if a temporary landowner carries out the land development. As in many of the cases, within the operation, cross-subsidising between profitable and unprofitable parts is possible.

In the course of the development process, the requirements regarding the level of servicing were sometimes changed by the municipality. When that happened, there was a renegotiation between the SEMAEB and the municipality. In cases where the costs of the new requirements were higher than what had been agreed initially, the municipality sometimes made an extra contribution. The municipality put money into the budget of the operation also, by buying the land required for public services such as schools or sports facilities. It acquired this land at a reduced price, but this generated income for the developer. However, this is not seen as a contribution by the municipality. On the contrary, the fact that the land was sold at a reduced price could be considered as a contribution from the developer towards the servicing of the area.

Depending on variables, such as the condition of the land, the housing density, the required level of servicing, etc., the costs for land development vary considerably throughout the cases. The costs found in the case of Rennes la Poterie, described above, are average. The lowest costs for land development, a little over 15 Euro per square metre, have been realised in the case of Bois-Guillaume Les Portes de la Forêt (table 3.8). This low cost can be explained by the relatively low density of the housing scheme (15 dwellings per hectare, the lowest of all the cases), by the fact that some of the costs that figure under the costs of land development in other cases, are considered here as a contribution towards secondary services, and by a deliberate – and creative – effort to keep the costs of land development low. This last point is illustrated by the hierarchical road design, but better still, by the system of on-site rain water treatment, which replaces more conventional methods of drainage.

The area of the Portes de la Forêt is situated on a plateau north of the city of Rouen. Normally, the rain water from this site would be drained into the river Seine. This would require installating drain pipes leading the water down from the plateau towards the river, involving high costs, since the distance that has to be covered and the difference in altitude between the plateau and the river are considerable. To avoid these high costs, the private land developer proposed a solution in which the rain water was dealt with 'on-site'. He argued that this solution had another advantage – besides the lower costs – in that it allowed increasing the residential quality by offering an attractive green space.

In the cases of Stuttgart Hausen-Fasangarten, Bonn Ippendorf, Cramlington North-East Sector, and Bishop's Cleeve, the costs for land development are considerably higher. Here they vary between 40 and 50 Euro per square metre. It is hard to say on the basis of the information available why these four cases show such high costs for land development. It must be related to the (required) level of servicing and to the variety of costs that are put under this heading on the financial account. We return to this in the last section of this chapter.

Contributions towards secondary services

Contributions towards secondary services were in some cases an explicit item on the financial accounts of the housing schemes. But what is seen as contributions towards secondary services is largely influenced by the context, more specifically by government regulation as to what services a developer of a site should provide. The difference in costs for land development (in other words: for realising the primary services) between the cases can partially be explained by differences in the services that are

112

considered as being secondary services, and hence taken out of the primary concerns of the developer. In the last case, they do not figure on the financial account as a part of the costs of land development. In some cases, 'contribution towards secondary services' figures as an item on the financial accounts. We have seen that in Zwolle Oldenelerbroek (table 3.3). It is also the case in Bois-Guillaume Portes de la Forêt (table 3.8).

In cases where the contribution by the developers towards secondary services does not appear explicitly on the financial accounts, this is hard to evaluate. This is for example the case in Bishop's Cleeve (table 3.4). In exchange for getting permission to develop a part of Bishop's Cleeve, the developers provided the municipality with 13 hectares of land to be used for community facilities. This land was situated outside the plan area where the housing development took place. In addition to this land, the house builders paid 770,000 Euro to the Borough Council towards the development of the area. Then, the house builders had to provide 2.2 hectares of land free of charge to the planning authority, to build a school in the area, and 0.8 hectares for a playing field. The value of this land is considerably higher than that of the area discussed above, because this was serviced land, which would otherwise have been developed as housing land. Finally, the house builders had to contribute to the financing of a bypass around the development site. Information about how much of the costs of the bypass they paid are not available. It is therefore not possible to calculate the financial margin received by the developers. In table 3.4, the figures that are available are presented. In total, we can conclude that the house builders received less than the financial margin mentioned in the table, because of the contributions they had to make to secondary services.

Interest costs

Whether or not interest costs play an important role depends upon the time the land has been held by an actor. When an actor acquires the land, he makes an investment. When he does this with borrowed money, it is evident that he pays interest costs to the lender. But even if the actor has been able to buy the land out of reserves, he pays interest costs on the investment. As long as the land is not yet developed, there are no returns on the invested capital. During this period, the buyer of the land misses the

113

potential income he could have gained by using the money now invested in the land. In other words, he pays interest costs.

It is interesting to note that a high level of interest costs often comes in combination with a low level of acquisition costs, and vice versa. This can easily be explained. We have seen that the costs of land acquisition are mainly determined by the level of 'hope value' that exists. If the land is bought very early, when there is not yet any development pressure or expectations that the land is going to be developed, the hope value is low. But in these cases, it usually takes quite some time before actual development takes place. All this time, the buyer of the land has to pay the interest costs. This can be observed in Rennes la Poterie (table 3.9), and Zwolle Oldenelerbroek (table 3.3), but it is most clearly illustrated by the case of Arnhem Rijkerswoerd (table 3.2).

The land in Arnhem Rijkerswoerd was bought a long time in advance of any development. At the time when the municipality acquired the land in Rijkerswoerd, at the beginning of the 1970s, it was used as farmland. The land had a value in this agricultural use of between two and three Euro per square metre. At the time the municipality acquired the land, there was no speculation. The municipality bought the land at its agricultural value. That was common practice in the Dutch housing development of the time. So the costs for land acquisition were not especially high. Nevertheless, they represented a considerable amount of money. The municipality financed this with a loan. Because it took a long time (almost twenty years) between the land acquisition and its development, the interest costs on this loan were considerable. Although there was rent received from renting the land to farmers before it was developed, and this partly covered the interest costs, the latter represent a substantial part of the total costs of the development.

In the cases where there is no intermediary landowner, i.e. in Stuttgart Hausen-Fasanengarten and in Bonn Ippendorf (tables 3.6 and 3.7), interest costs do not play a role since there is no actor who invests in land acquisition. In all the other cases, interest costs did occur. However, they are not always explicitly mentioned in the financial accounts. Especially in the cases of Cramlington North-East Sector (table 3.5) and of Bishop's Cleeve (table 3.4), it was not possible to bring them to light. The interest costs might be incorporated in the costs for land development, which would explain the high level of these costs in those cases.

114

The 'regular profits' that are mentioned in the scheme require some explanation. As mentioned above, these profits are directly related to the risks and uncertainties the actors take in the process. The regular profits directly influence the possible occurrence of a financial margin in the process. Therefore, there is also a direct link between the risks and uncertainties in the process, and the possible financial margin. It is therefore important to consider the risks and uncertainties that might appear in more detail. Uncertainty in the following variables plays a role when a developer evaluates whether or not to invest in housing development:

- price of land (e.g. an unexpected case of pollution that necessitates site decontamination);
- interest level (e.g. unexpected changes in the financial markets);
- number/type of houses to be built (e.g. modification of the plan in the course of the development process);
- construction costs (e.g. a sudden price rise of building materials);
- time of construction (e.g. a delay in the delivery of required building materials);
- revenues from disposal or rent (e.g. a change in the market for housing);
- required contribution towards the residential environment (e.g. when a municipality decides in the course of the process to change its requirements).

During the development process, these variables can change. To the developer, this means that there are risks which are the result of uncertainty about time and costs of each of the activities. Not all of the possible uncertainties are important in each process. The uncertainties depend on the specific circumstances in each process. They also vary in importance between different processes. To compensate for these uncertainties, the actors in the process will require profits. If the uncertainties are not compensated by profits, the actors concerned will not want to take the risks. We call these profits the regular profits, and because they are necessary to allow the process to continue, they can be considered as costs.

Only in the case of Bois-Guillaume Portes de la Forêt (table 3.8) and Rennes la Poterie (table 3.9), do the regular profits appear explicitly on the financial accounts of the operation. This is most explicit in the case of Bois-Guillaume Portes de la Forêt, where they can be found under this heading on the financial account with 1.83 Euro per square metre. Besides, there is a reserve for unexpected expenses, at 0.28 Euro per square metre, which can also be seen as a form of regular profits if the unexpected expenses do not occur. In Rennes, the regular profits are presented as part of the development costs (2.29 Euro per square metre). In the other cases, regular profits are probably incorporated in the other items. As the description above shows, the regular profits vary according to the characteristics of each development. It is therefore not possible to say what would be a reasonable level of regular profits in each of the cases.

3.4 The income side

In much the same way as the previous section dealt with the costs in the process of land conversion, this section deals with the income side of the financial accounts. The items on the income side are less varied than those on the cost side. Basically, the only income during the process of land conversion comes from the sale of either unserviced or serviced building land. A supplementary source of income might come from subsidies, either from higher levels of government, or from other housing schemes within the same municipality, which were financially more successful. But in whatever form, subsidy is always an input from the outside into the development process, whereas the income from the sale of land is generated within the process.

Sale of unserviced building land

The sale of unserviced building land represents both income for the first landowner, and costs for the developer or the temporary landowner. The income from the sale of building land is a very important item on the

116

financial account, because it determines the financial margin in the rest of the process. In the case of Stuttgart Hausen-Fasanengarten, the high value of the building land (both serviced and unserviced) is the cause of a considerable financial margin, on the market for both unserviced and serviced building land. As in Bonn Ippendorf, there was no temporary landowner who took care of the development. The first landowners remained in posession of the land and were responsible for the servicing of the area. As a result, the first landowners received an important part of both financial margins. The municipality of Stuttgart created a mechanism to use this financial margin to cover some of the costs of the development of the area: the *Umlegung mit freiwillig vereinbarten Konditionen* (land reallocation under voluntarily agreed conditions). For the way in which the financial analysis is carried out in this case without a temporary landowner, we refer to the example of Bonn Ippendorf in table 3.6.

Whereas the first landowners in Stuttgart Hausen-Fasanengarten received a high income as a result of the high price of the building land, the landowners in for example Cramlington North East Sector (table 3.5), Rennes la Poterie (table 3.9), or Arnhem Rijkerswoerd (table 3.2) did not have such a high income. That means that in these cases, the first landowners did not receive a high financial margin on the market for unscrviced building land, especially when also regular profits that can be accredited to them are taken into consideration. Where a financial margin appeared in these cases, it was almost entirely received by the developer as a margin on the market for serviced building land.

Disposal of serviced plots

The disposal of serviced building plots generates the major part of the income during the process of land conversion. Normally, it is the intention of the developer that the disposal of the serviced building plots covers the costs of land acquisition and of land development, i.e. of the land conversion. If the disposal of the serviced plots yields more than the costs of the land conversion, the difference is a financial margin on the market for serviced building land. The disposal price of the plots is dictated by the market. Usually it is not possible to raise the price of the plots, for example by loading extra costs on them. If the price for building plots becomes too high, no one will buy them. Potential buyers will search for alternatives elsewhere. Therefore, in the end it is the market situation that determines the financial margin in a development process, in combination with the level of costs.

For example, in Arnhem Rijkerswoerd (table 3.2), the interest costs, together with the costs of the land development, took a large part of the income from selling serviced land. This price, at which the municipality disposed of the building plots, should cover all the costs of the development. But as mentioned above, that price was dictated by the market: if the municipality had set the price too high, nobody would have acquired the land because it would not have been economically viable to build houses on it.

At this market price, the income from selling the building plots was not enough to cover all these costs. A central state subsidy (*locatiesubsidie*) was required to bridge the gap. This also means that there was no financial margin on the market for serviced building plots in Arnhem Rijkerswoerd. Nevertheless, we use the notion of financial margin here, to understand how the expenditure on the residential environment was arranged in this development process. In this case, the municipality had – implicitly – made a similar calculation beforehand: the municipality adapted the level of servicing of the area to the money that would be generated by the sale of the serviced plots and the subsidy.

Price discrimination between plots for different types of housing can be found in most of the studied housing schemes. We take as an example Bois-Guillaume Portes de la Forêt (table 3.8). According to the type of dwelling, the plot price differed. Five types of housing were distinguished in the plan: 80 *lots libres* (free building plots) with a plot price of 44,850 Euro, 70 *maisons individuelles groupées* (grouped individual houses) with a plot price of 35,1000 Euro, 30 *maisons de ville* (terraced houses) with a plot price of 17,250 Euro, 125 *logements collectifs accession* (apartments for sale) at 13,500 Euro per plot, and 50 *logements collectifs locatifs* (rented apartments) at 6,000 Euro per plot. The differences in plot prices are due to the difference in size between the plots and a different price per square metre for the different types of plots. This makes it complicated to calculate the price of serviced land in the area. The average price paid for a plot (all types of housing together) was about 24,000 Euro.

In the cases where the land development and the housing development are carried out by the same actor, there are usually no serviced building plots disposed of. That can be seen in the cases of Cramlington North-East Sector (table 3.5) and Bishop's Cleeve (table 3.4). Instead of selling the serviced building plots, the developers in both Cramlington North-East Sector and in Bishop's Cleeve bought unserviced land, constructed houses on that land and sold them. The process of land conversion is thus extended and includes also the construction of the houses. The temporary landowner responsible for the land development is at the same time the builder of the houses on the site. For our analysis, this means that in these cases it is not possible to distinguish a financial margin on the market for serviced building land separately from a financial margin on the market for new housing. The costs of housing construction must be added to the costs of the land conversion, so the total costs incurred by the developers are higher. But they receive income from selling houses instead of serviced plots, so the income is also higher.

Subsidies

The case in which additional income from subsidies played the most important role was in Arnhem Rijkerswoerd (table 3.2). Here the municipality carried out all the servicing, both primary and secondary. The costs of servicing were then passed on to the developers by incorporating them into the plot prices. However, the municipality could not put all the costs of servicing into the plot prices because the latter could not be higher than the market prices, or than the fixed plot prices (social housing). That is why central state subsidy (in the form of *locatiesubsidie*) was required for the realisation of Rijkerswoerd. This way, the municipality managed to balance costs and income.

The municipality had high requirements with regard to what it saw as residential quality. So the costs of the development were high. When the income generated by the land development appeared to be insufficient to cover all the costs, central government helped with *locatiesubsidie*. This also implied that central government fixed standards for the residential quality, or at least agreed with those fixed by the municipality. Without the help of central government it would not have been possible to realise these standards of residential quality.

The positive balance in Zwolle Oldenelerbroek (table 3.3) was used to cover negative results in other housing schemes carried out by the municipality of Zwolle. In these other schemes, this money was an input from outside, hence we can consider it as a subsidy. In Arnhem Rijkerswoerd (table 3.2) this worked the other way round. The negative result here was covered by positive results in other housing schemes for which the municipality of Arnhem carried responsibility. However, this cannot be seen on the financial accounts of the operation because the loss on Rijkerswoerd Phase Two was transferred to the following phases of Rijkerswoerd. In consequence, these phases started with a negative balance. Other forms of subsidy might exist, but they did not occur explicitly in our cases. Using money gained by selling expensive building plots to finance affordable building plots within the same housing scheme is sometimes referred to as cross-subsidising. However, as it is not an input from the outside into the process, we do not consider this as subsidy. It is a way to use the money generated in the process to increase the residential quality of the housing scheme, because when cross-subsidising occurs, a mix of tenure is usually seen as a feature that increases the quality of the scheme. This argument is sensitive to the delimitation of the plan area. The larger the area, the greater the possibilities for cross-subsidising.

3.5   Expenditure on the residential environment

Above, data from the different cases have been used to show how the financial accounts of housing development processes can be analysed. The financial data were presented per item on the financial accounts. As described in section 1.5, the central element in our analytical framework is the type of development process, determined by which actor is the (temporary) landowner at the time of the land development and is therefore responsible for the land conversion. In this section, the information presented above is restructured using this analytical framework. This makes the link with the next chapter, in which the analysis focuses on the factors that determine the way in which the money is used. The necessary first step in that analysis is seeing who received what, and who paid what. The financial analysis in this chapter allows us to distinguish that.

The incidence of the costs, of both primary and secondary services, is shown in summary form in table 3.10. The provider of the services is also shown, as it is the provider who runs the risk of not being able to recoup the costs. Thus, this table can be seen as a recapitulation of the previous sections. It is used here to describe the relationship between the type of development process and the incidence and size of a financial margin. Thus it provides the link between this chapter in which a financial analysis of housing development processes is carried out, and the next chapter, in which the focus is on the institutional aspects of these processes.

Table 3.10   Incidence of costs and gains in five types of development processes

Type of process	Primary services	Secondary services	Financial margin
I Public body as temporary landowner	Provider: - Municipality Method of payment: - Included in price of plots	Provider: - Municipality Method of payment: -Included in price of plots - Otherwise: taxpayer	Received by: - Municipality - Original landowners
II Public-private body as temporary landowner	Provider: - (Semi-) public body Method of payment: - Included in price of plots	Provider: - Municipality Method of payment: - Included in price of plots - Otherwise: taxpayer	Received by: - (Semi-) public body - Original landowners
III Private body as temporary landowner	Provider: - Private body Method of payment: - Included in price of plots	Provider: - Municipality Method of payment: - Out of planning gain - Otherwise: taxpayer	Received by: - Private developer - Original landowners
IV No temporary landowner, no public powers	Provider: - Public body (probably municipality) Method of payment: - Partly recovered from landowners (agreement) - Otherwise, taxpayer	Provider: - Public body (usually municipality) Method of payment: - Possibly planning gain agreement - Otherwise: taxpayer	Received by: - Original landowners
V No temporary landowner, use of public powers	Provider: - Public body (probably municipality) Method of payment: - Partly recovered from landowners (depending on public powers) - Otherwise: taxpayer	Provider: - Public body (usually municipality) Method of payment: - Depends on the content of the public powers (development obligation, levy, ...) - Remaining costs by taxpayer	Received by: - Original landowners

When this table is filled in with the information presented in this chapter, this gives the following results:

1   Public body as a temporary landowner (Zwolle Oldenelerbroek, Arnhem Rijkerswoerd)

In these cases the private (house builders) or semi-public (housing corporations) participants in the process were responsible for the housing construction only. The municipality took care of the land assembly, the land development and both primary and secondary servicing of the area. In the case of Arnhem Rijkerswoerd, a subsidy from central government was required to make this financially possible. So in this case, a negative financial margin on the market for serviced building land of almost 10 million Euro (the size of the subsidy) was borne by central government. And after that, a negative financial margin of 815,000 Euro remained for the municipality. The fact that the municipality had paid only the agricultural value for the acquisition of the unserviced building land could not prevent this, and the first landowners did not receive a financial margin. This must partly be explained by the high interest costs, due to the long period the municipality was owner of the land before being able to develop it. In Zwolle Oldenelerbroek, a central state subsidy was not necessary. The fact that the land was acquired at about four times its agricultural value resulted in a financial margin on the market for unserviced building land of 2.86 million Euro (from which the regular profits need to be subtracted). This relatively low price of acquisition, together with costs of land development, and interest costs which were lower than in Arnhem Rijkerswoerd meant that the municipality received a financial margin on the market for serviced building land of 1.9 million Euro. The taxpayer did not have to contribute to the development of Zwolle Oldenelerbroek, in effect, the operation financed itself. In Arnhem Rijkerswoerd, through the subsidy from central government, a substantial contribution from the taxpayer was required to realise the housing scheme.

2   Public-private body as a temporary landowner (Rennes la Poterie)

This case is very much like the first two cases in which the municipality (a public body) acts as a temporary landowner. This is even more so because

123

in the case of Rennes la Poterie, the municipality had followed an active policy of land acquisition, as a result of which the public-private developer could acquire it for a fairly low price (4.45 Euro, a little over four times the agricultural value). The price at which the municipality sold the land to the developer consisted of the costs of acquisition of the land from the first landowners plus the interest costs that the municipality had incurred. Another reason for the similarity between the cases where a public body is temporary landowner and the case where a public-private body is temporary landowner is that neither of these parties have to make profits to ensure their continuation. Therefore their reasoning is rather similar.

Generally, the costs for primary as well as secondary servicing had to be covered by the disposal of the plots. To this aim, agreements were made between the municipality and the developer as to the contributions towards the secondary servicing. The municipality – hence the taxpayer – also made some contribution, to balance the financial account.

3 Private body as a temporary landowner (Cramlington North-East Sector, Bishop's Cleeve, Bois-Guillaume Portes de la Forêt)

There is a difference between the cases of Cramlington and Bishop's Cleeve on one side, and the case of Bois-Guillaume on the other side. In the first two cases, the land developer also built the houses on the site and then sold them. In Bois-Guillaume, the land developer sold serviced building plots to the house builders. This has as a consequence for the financial margin on the market for serviced building land, which can be distinguished in the case of Bois-Guillaume Portes de la Forêt, but cannot be found in the cases of Cramlington and Bishop's Cleeve. In the last two cases, it is combined with a financial margin on the market for new housing. However, this does not affect who pays what, and who receives what.

In all three cases the developers took care of the primary servicing, and in all cases there were arrangements to make the developers contribute to the secondary services. The local planning authority's role was one of negotiation about what was to be realised, and of supervision. Moreover, the municipality was responsible for the provision of part of the secondary services. In return, it received contributions from the developers through the construction of the bypass in Bishop's Cleeve, for example, or the flyover in Bois-Guillaume Portes de la Forêt. These contributions are based upon an estimation of the extent to which the facility serves the new residents of the housing scheme, and the extent to which it serves others. It can also be agreed that a developer takes the responsibility for the realisation of secondary services, which is a contribution in kind rather than in money.

4   No intermediary landowner, no public powers (Stuttgart Hausen-Fasanengarten)

The case of Stuttgart Hausen-Fasanengarten is a special one in many respects. First of all, because of the very high land prices that occurred here, 229.50 Euro per square metre of unserviced, and 450 Euro per square metre of serviced building plots. These are the highest in the cases that have been studied. The high land prices have incited the municipality to find a creative solution in order to be able to realise some of its quality criteria, especially the construction of social housing in the area. The *Umlegung mit freiwillig vereinbarten Konditionen* allowed the municipality to impose development obligations upon the developers. This means that the developers, who were in this case also the original landowners, paid for a part of the secondary services. But another part of the secondary services, for an amount of 4.2 million Euro, remained to be funded by the municipality, hence to be paid ultimately by the taxpayer.

Besides this contribution towards secondary services, the developers paid for all of the primary services. This is unusual in Germany, where the most commonly used procedure (the *Erschließungsbeiträge*) allows the municipality to recoup only 90% of the costs of land development from the developer. As a result of the huge value increase of the land due to the designation as housing land, a financial margin remained. Because the first landowners were also the developers, and they disposed of the land only after it was developed, as serviced building plots, the distinction between a financial margin on the market for unserviced and for serviced building land does not apply here. The total financial margin, received by the landowners/developers amounted to 22.9 million Euro.

In this case, strict regulations applied as to who paid what and who received what. The landowners had to provide the land required for the primary servicing of the area (mainly for the infrastructure). In Bonn Ippendorf, this meant that they had to provide 17.2% of their land free of charge. But in any case, the municipality is entitled to receive the value increase of the land due to the servicing. That meant that in Bonn Ippendorf, it received in addition to the land another 1.44 million Euro. This money was used for the secondary servicing of the area. Costs for primary services were recouped from the landowners by another mechanism, the *Erschließungsbeiträge* (contribution to land development), which allows the municipality to recoup up to 90% of these costs.

The value increase of the land from agricultural to building land goes to the first landowners. This value increase was big enough so that after having paid their *Erschließungsbeiträge*, the landowners still received a financial margin on the market for unserviced building land (although they did not actually receive it since they kept the land in their hands). The value increase between unserviced and serviced building land went to the municipality. After having subtracted all the costs (for planning, administration, and secondary servicing) the municipality received a financial margin of 765,000 Euro.

4 Actors and activities, roles and relations

4.1 An institutional analysis of the development process

After the analysis of the financial aspects of the process of housing development in chapter four, we now turn to the institutional aspects of that process. Which actor plays which role? How does this affect his strategy? What are the relations between the actors? This chapter answers these questions for the eight case studies.

The development process according to Chambert and Edwards

As said before, the greenfield housing development process is defined as the process during which a site is transformed from its original, non-building use to a housing use. For this transformation, a number of activities have to be carried out, e.g. land assembly, land development, housing construction, marketing the end product (see Healey, 1992b). These activities can vary as to the order in which they are carried out. Nevertheless, Chambert (1988) and Edwards (1995) claim that it is possible to distinguish four stages in the development process. These should not necessarily be seen as sequential steps, but as the constituent parts of the process.

Chambert and Edwards use the distinction of different stages to visualise the actors, the roles and the interactions in the development process. This schematic representation is a good tool to analyse the difference between development processes (see also Magalhães, 1999). Although the schematic representation of the development process that is used in this section originally comes from Chambert (1988), in our representation we refer mainly to Edwards (1995). The latter has used Chambert's framework for a concise discussion of agents and functions in urban development. He distinguishes four main stages that can be recognised in the development process. These are the previous land use, a mediating land ownership, the production, and the use or consumption of the end product. In a housing development process, all these stages necessarily have to be passed.

During each of these stages, a number of functions have to be fulfilled. Edwards speaks purposely of 'functions' rather than of 'agents' or 'players'. He argues that: 'The main functions in urban development always have to be performed, while particular kinds of agents may be absent in a particular case.' Moreover: 'The functions may be performed by a single agent or by a lot of separate ones and it is useful to look at these different organisational forms as important in themselves' (1995: 26). The main functions both Chambert and Edwards distinguish are:

–   finance: all the stages of the development process that are distinguished involve financing. The degree of influence the financiers and the lenders have on the process can vary considerably from case to case. Insight into this function is required to understand why and how the events in the development process occur;
–   production: a particularity of the process of house building is that it often – but not necessarily – involves separate designers and builders. Further, what is meant by this function is clear;
–   ownership/promotion: as opposed to production, the way in which this function is organised and realised varies greatly from one case to another. It can be very simple, as where a municipality builds its own school on its own land. But it can also be a complex function that requires a lot of coordination between different actors and activities;
–   use: there are several forms for the use of real estate. Owner occupation is the simplest, the alternative is a range of forms of tenancy.

By putting these functions and the stages in a matrix, we can conceptualise the development process as a number of functions that are fulfilled in a number of stages. The resulting diagram (see table 4.1) allows to show which actor fulfills which function in which stage of the process: the sixteen boxes are functions being performed by an actor in a particular stage of the process. There may be more than one actor in a box, and there may also be some empty boxes when at a certain stage no one carries out that function (e.g. during the stage of the previous land use, the function of production was not performed).

Table 4.1   The development process according to Chambert

phase ⇒ ⇓ function	previous land use	mediating land ownership	production	ownership/use (consumption)
finance – credit				
production				
ownership – promotion				
use				

Source: Chambert, 1988

Edwards stresses that what is often important are the relations between the actors. These can take varied forms, such as contracts for work, tenancy relations, interest on borrowed money, or outright purchases and sales. The general conclusion Edwards draws from the discussion of Chambert's model of the development process is that: '... we have to try to understand the interactions of the system as a whole' (1995: 37). But also that: '... the state (central and local government and all the public bodies) are centrally important in constituting and regulating all the relationships and functions shown in the Chambert table' (1995:38).

Translation into this study's terminology

The above framework for the analysis of the development process was the starting point for our analysis, but it has been adapted to suit the aims of this study. There are two reasons for this. First, we do not need a framework that applies to all real estate development processes. We are interested only in one of them, i.e. the development of greenfield housing.

That means that the framework can be adapted to precisely that type of development process. Further, we aim at highlighting other aspects of the development process than Chambert and Edwards do.

First, the stage of production is separated in a stage of land development, and a stage of housing production. The reason for this is the central role that is accorded here to the land assembly and land development. That makes it necessary to distinguish this function separately. Furthermore, the notion of 'functions' is replaced by the notion of 'roles', and the notion of 'stages' by the notion of 'activities'. The reason for preferring 'activities' to 'stages' is that the latter implies a chronological order that we do not want to imply. Speaking of the activities that take place in a development process is more neutral as regards the sequence in which they are realised. In practice, in an area to be developed, different activities are often carried out simultaneously, for instance when houses are being built somewhere in the area while the land assembly for the whole area is not yet finished.

The reason for replacing 'functions' with 'roles' is mainly textual. In institutional analyses different words are used for this concept. Ostrom (1986) speaks about the 'positions' of participants in a process, Healey (1992b) of roles. We find that – in combination with the notions of actors and activities – 'roles' gives a clear idea of what is meant here. Because we speak of roles instead of functions, we formulate them differently. Thus, the function of land ownership becomes the role of landowner, the function of production becomes the role of producer, etc.

But our adaptations of Chambert's and Edwards' scheme go further. In the first case studies we found that the functions that they distinguish are not all of great importance in our perspective. On the other hand, some other functions (or in our terms: roles) were important in order to understand the influence of the housing development process on the residential environment. Thus, the function of finance and of end use have not been investigated separately in this study. Conversely, the function of ownership-promotion has been separated into the role of owner, and the role of promoter, or in our terms, of initiator. The role of the designer has been introduced because this allows an important aspect of the residential environment, i.e. the urban design, to be analysed.

With the above amendments, Chambert's model of the development process can now be represented in table 4.2. In this form it is used in this study to analyse the development process of the eight cases. We are aware that by choosing such a format before doing the fieldwork, the viewpoint of the researcher has already been chosen. However, as argued in chapter two, the choice to enter the case studies with a clearly defined analytical framework has been made deliberately. Moreover, this framework is flexible enough to include very different forms of development processes.

Table 4.2   Activities, roles, and actors in the development process of greenfield housing

activities ⇒ ⇓ roles	identification of opportunities	land assembly	land development	housing production	owner-ship/use
owner					
initiator (or: promotor)					
producer					
designer					

As said before, in this study a key role is accorded to the actor who is responsible for the land assembly and the land development: the (temporary) landowner during land development. When using the framework above to describe the actors, roles and activities in the process, it shows clearly which actor took on this role. Distinguishing the actors and their roles in this way makes it possible as a next step to analyse the strategies and interests of each of the actors. Moreover, the framework can be used to describe the relations between the actors in the analysed development process. Thus, an important part of Healey's (1992b) four levels, through which an institutional analysis of the development process should proceed, are covered. This opens the way for an important part of the analysis that has not yet been considered, i.e. the analysis of power and dependence and the way in which this influences the outcomes of the

131

process. We deal with that in section 4.4. First, in sections 4.2 and 4.3 the framework described above is applied to the case studies.

4.2 Case study files: actors, roles, and activities

Arnhem Rijkerswoerd

In the development of Arnhem Rijkerswoerd, the municipality of Arnhem played a central role. It bought the unserviced building land when this was still in agricultural use. It then carried out the land development and eventually sold the land as serviced building plots to private house builders and to housing corporations. The municipality carried out the land development itself. So it was able to realise a level of services and a parcellation that corresponded to its own ideas about the residential environment. It actively constructed the residential environment. After this, it sold the building plots only, all the public spaces and the roads remained in the hands of the municipality.

Central government played an important role in the development process: Without the *locatiesubsidie*, it would not have been possible to realise the housing scheme. To grant this subsidy, central government imposed requirements concerning the form of the development. If it had appeared later that the housing scheme once realised did not comply with what was agreed upon, the money would have been claimed back.

The private house builders primarily wanted to build houses and earn income from that activity. In Rijkerswoerd, they had to make their profit on the construction and sale of the houses, since the municipality carried out the land assembly and the land development. The role of the private house builders in the development process was to finance and construct houses, within the limits set by the municipality. Private house builders and municipality were partners in the last phase of the development process, the housing construction. The municipality could not impose unrealistic demands from a 'marketing' point of view, since house builders closely follow the market, to see what they can and what they cannot build and sell.

Table 4.3   **Activities, roles, and actors in Arnhem Rijkerswoerd and Zwolle Oldenelerbroek**

activities ⇒ ⇓ roles	identification of opportunities	land assembly	land development	housing production	ownership/ use
owner	- agriculturalists	- n.a.	- municipality	- house builders - housing corporations	- owner-occupiers - institutional investors - housing corporations
initiator	- municipality	- municipality	- municipality	- house builders - housing corporations - municipality	- n.a.
producer	- n.a.	- municipality	- municipality - sub-contractors	- house builders - housing corporations - sub-contractors	- n.a.
designer	- n.a.	- n.a.	- municipality	- house builders - housing corporations - municipality	- n.a.

The housing corporations also had an interest in the residential environment since they were concerned not only with finding a first occupant for the house. The dwellings they built remained their property, so they were also concerned with long term maintenance, and with the question of whether they would still be able to find occupants in the future. Their interest in the long term viability of the housing scheme made the housing corporations more concerned with the residential environment.

Figure 4.1   A street in Arnhem Rijkerswoerd

Zwolle Oldenelerbroek

The actors who played a role in the development of Zwolle Oldenelerbroek were the municipality of Zwolle, a number of private house builders, three housing corporations and the original landowners. The division of roles was identical to that in Arnhem Rijkerswoerd, see table 4.3.

A large part of the activities in the process were directly carried out by the municipality, but even the activities carried out by other actors were largely influenced by the municipality. Firstly, because the municipality set the limits of the development in the local plan. Secondly, because the municipality as landowner could choose its (private) partners. The private agreements coupled to the sale of the land were another opportunity to fix specific conditions for the development of the site. The coordination between the departments within the municipality was assured by involving all the departments from the beginning and by synchronising the different planning procedures (i.e. infrastructure, facilities, housing) within the framework of the local plan procedure.

In addition to the municipality, the private house builders played an important role. There were no difficulties with getting house builders interested in building in Oldenelerbroek. The market was good, so the house builders were happy to be able to build in this location. Apparently, the close supervision by the municipality was not a problem for them. Of course the house builders made their own calculations to find out whether it was economically interesting to build, and they tried to give their own 'touch' to the design of the houses. That was the only aspect in which they could express their own vision of the development, and thus interest possible buyers.

The housing corporations had other objectives than the private house builders. Like the municipality, they aimed at balancing costs and income. The houses they built remained their property, they wanted to be able to let them for a long time. Like the private house builders, the housing corporations were responsible only for the design of the houses, within limits fixed by the municipality. As a result of a partial privatisation of the housing corporations in the beginning of the 1990s, the corporations started to act more like market-driven parties. As a first sign of this, a housing corporation built some private sector houses in Oldenelerbroek.

The first landowners did not play an active role in Oldenelerbroek. But in a way, the role they played was very important. Because they sold their land to the municipality without much delay, the latter was able to ensure the continuity of the development. It did not have to fall back on time-consuming expropriation procedures. Higher tiers of government played a direct role in the development of Oldenelerbroek only through the planning law: they checked and approved the local plans. Earlier in the development of Zwolle Zuid, subsidies had been paid for the realisation of large infrastructure works, but these were not directly part of Oldenelerbroek.

Bishop's Cleeve

The different parties that played a role in Bishop's Cleeve were Tewkesbury Borough Council, Gloucestershire County Council, and two private house building companies that owned the land in the area. Other house builders who were active on the site had to comply with the agreements signed between the two landowning house builders and the local planning authority. Their role in the realisation of the housing and thus of the residential environment was therefore only marginal. The first landowners did not play an active role. The land was acquired early in the process. It was already fairly clear that housing development was going to take place, so the landowners did receive some of the financial margin.

Table 4.4   Activities, roles, and actors in Bishop's Cleeve

activities ⇒ ⇓ roles	identification of opportunities	land assembly	land develop- ment	housing production	ownership/ use
owner	- agricultur- alists - house builders	- n.a.	- house builders	- house builders	- owner- occupier
initiator	- county council	- house builders	- house builders - borough council	- house builders	- n.a.
producer	- n.a.	- house builders	- house builders - sub- contractors	- house builders - sub- contractors	- n.a.
designer	- n.a.	- n.a.	- house builders - borough council - county council	- house builders	- n.a.

Following the proposals by the County Council, which marked the start of the process, the Borough Council picked up the development process and worked it out further. This was a statutory duty because of the Borough Council's responsibility for making a local plan in which future land uses were designated. This was done in the form of the Cheltenham Environs Local Plan which formed the basis of the development. The main objective of the Borough Council was to provide a residential area that satisfied the residents. To pursue this objective, the Borough Council used its competences for development control: It supervised the development, and had through the planning permission a statutory power to oblige developers to comply with the standards it set. An actor who also played a role in the development was the Parish Council. The reason why it is not represented in the table is that its influence went through the Borough Council.

Figure 4.2   A hint at traditional architecture in Bishop's Cleeve

All the County Council did was to allocate in the structure plan 1000 houses to Bishop's Cleeve. The way in which they were to be realised was decided by the Borough Council in the local plan and by the house builders who actually filled in the broad outlines given in the local plan. The objective of the County Council at this stage of the development was to accommodate the necessary number of houses in the area. Later on in the development process, the role of the County Council was that of 'highway authority'. In this role, the County was responsible for the layout and the finishing of the roads. Its objective was in the first place to create a safe road network. Another responsibility of the County Council was to make sure that enough schools were provided for (future) inhabitants of the housing scheme.

The house builders carried out the actual development, within the limits set by the local planning authorities. The main objective of the house builders was to generate an income by building and selling houses. Their main income does not come from the construction of the houses, but from the 'turnover' of the land, i.e. the value increase of the land between the moment it is bought and the moment it is sold. The house builders tend to build houses only when they know they are going to sell them, when a buyer has signed for a house. So the market demand largely influences how many houses are going to be developed in a certain period, and which types. This is reflected in the residential environment, where areas of development of the same period can be distinguished. As part of their objectives, house builders respond to changes in the market, even though these might include deviations from the plan. In Bishop's Cleeve, this resulted in higher housing densities than proposed by the District Council. After the development, the house builders left the area.

Cramlington North-East Sector

In the development process of Cramlington North-East Sector, private house builders were responsible for the process of land conversion. After that, they also took care of the construction of the houses. This gave them a central place in the development process. They carried out the major part of the activities.

Two private housebuilding companies were active in Cramlington North East Sector. Their main objective was to sell houses. The residential environment was important to the house builders in so far as it influenced people's perception of an area, which should be reflected by the house prices.

Northumberland County Council played quite an important role in initiating the development of Cramlington as a new town. Because the District Council at the time was not big enough to deal with the development control of such a large development, the County Council offered some officers to assist the District Council. The County Council officers were gradually withdrawn as the population of the district grew, and hence the District Council grew in importance. The County Council retained its influence on two major points. It was the local authority responsible for the planning of roads and schools.

Blyth Valley District Council was responsible for development control. That meant that they supervised the whole development and decided – as far as their competences permit – what development could take place in the area, and in which way. The District Council could also play a role as an initiator of housing development, however, in Cramlington it is hard to trace who played the role of the initiator. The house builders bought the land a long time ago and it is not possible now to find out whether it was they who suggested to the District Council to develop there, or whether it was the District Council who had the idea first.

Table 4.5   Activities, roles, and actors in Cramlington North-East Sector

activities ⇒ ⇓ roles	identification of opportunities	land assembly	land development	housing production	ownership/ use
owner	- agricultur-alists - house builders	n.a.	- house builders	- house builders	- owner-occupiers
initiator	- house builders - county council	- house builders - district council	- house builders - district council - county council	- house builders	- n.a.
producer	- n.a.	- house builders	- house builders - sub-contractors	- house builders - sub-contractors	- n.a.
designer	- n.a.	- n.a.	- house builders - district council - county council	- n.a.	- n.a.

Bonn Ippendorf

In Bonn Ippendorf, there was no temporary landowner responsible for the land conversion. Two actors can be distinguished, the landowners (there were over 80 of them on the site) and the city of Bonn.

Of the municipal departments involved in the development process, the *Planungsamt* (planning department) issued the *Flächennützungsplan* (structure plan) that designated future land uses. Before the actual development, a legally binding local land use plan had to be drawn up. Therefore, the *Planungsamt* did not play a very big role in the actual development process. Its objectives were on a higher level, namely a balanced growth of the city of Bonn as a whole. The *Tiefbauamt* (department of 'physical development') carried out the land development and specified the technical norms to which the land use plan had to adhere. The *Kataster und Vermessungsamt* (department of land registry) prepared the land use plan, and it took the necessary measures to realise the plan in the *Umlegung*. This made the *Kataster und Vermessungsamt* the central actor in the development process. But in spite of this central role, it could not just do as it pleased. It had to conform to three types of guidelines, those issued by the *Bundesregierung* (federal government), by the *Landesregierung* (regional government), and those by other departments within the municipality itself.

There were about 80 different landowners in the area, of whom the municipality was the biggest. The *Umlegung* was imposed on these landowners, they could not choose not to take part in it. But that was not much of a problem to the landowners, because they knew that in the end they would receive a plot with a higher value than that of the plot they owned before the *Umlegung*. The objectives of the landowners varied. Some of them only wanted to receive the value increase caused by the land conversion. These owners generally sold their plot before the procedure of the *Umlegung* started, so they were no longer involved in the process. Others wanted to build a house on their new plot. These owners were eager to receive a plot that was well situated. Still others tried to speculate (to the extent that this was possible within the procedure) with their plot. They kept it during the procedure of the *Umlegung*, and hoped that during the procedure the value of the plots would increase (the values used for the calculations in the *Umlegung* are those at the beginning of the procedure). After the procedure, these owners sold their serviced plot to others who wanted to build houses. Of course, they were keen on getting building plots in good locations, because these were likely to yield a higher price.

Table 4.6   Activities, roles, and actors in Bonn Ippendorf

activities ⇒ ⇓ roles	identification of opportunities	land readjustment	land development	housing production	ownership/ use
owner	- (agricultural) first landowners	- first landowners	- first landowners	- first landowners - house builders	- first landowners - owner-occupiers
initiator	- municipality	- municipality	- municipality	- first landowners - house builders	- n.a.
producer	- n.a.	- municipality	- municipality - sub-contractors	- first landowners - house builders - sub-contractors	- n.a.
designer	- n.a.	- municipality	- municipality	- first landowners - house builders	- n.a.

Stuttgart Hausen-Fasanengarten

Three main actors were involved in the development of Hausen-Fasanengarten. These were the municipality of Stuttgart, the landowners in the area concerned, and the *Gesellschaft für Stadt und Landesentwicklung MBH* (GSL, society for for urban and rural development Ltd.).

The municipality of Stuttgart initiated the process. As a political actor, the municipal council took decisions about the course of the development process and the form of the housing scheme. For practical reasons, these decisions were taken not by the municipal council as a whole, but by a group of representatives whom the council installed to deal with aspects of land development and land assembly: the *Ausschuß für Bodenordnung* (committee for land development). This committee consisted of the mayor, who was the chairman, and 14 municipal councillors. Three technicians from the municipal administration took part in the meetings, but did not have a right to vote. Decisions taken in the committee were not made public. Only the decision to start an *Umlegung*, and the final presentation of the *Umlegungsplan* were made public.

The technical department of the municipality that was most concerned with the housing development was the *Stadtplanungsamt* (department of city planning). It made the development plan. In this plan, requirements as to the servicing of the area, the amount of open space, and the facilities to be realised were fixed.

Table 4.7   Activities, roles, and actors in Stuttgart Hausen-Fasanengarten

activities ⇒ ⇓ roles	identification of opportunities	land readjustment	land development	housing production	ownership/ use
owner	- first landowners	- first landowners	- first landowners	- first landowners - house builders	- first landowners - house builders
initiator	- municipality	- municipality - GSL	- GSL - first landowners	- first landowners - house builders	- n.a.
producer	- n.a.	- municipality - GSL	- first landowners	- first landowners - house builders - sub-contractors	- n.a.
designer	- n.a.	- GSL - municipality - first landowners	- municipality - GSL	- first landowners - house builders - municipality	- n.a.

The *Gesellschaft für Stadt und Landesplanung* (GSL) acted as an intermediary in the procedure. It took care of the supervision and management of the project. The designing and construction work was done by sub-contractors. The GSL represented the municipality in its dealings with the landowners, and the landowners in their dealings with the municipality. It negotiated with the landowners about the reparcellation of the area, and about their contribution to the land development. It then presented the municipality with a plan that all the landowners agreed upon. This took a lot of work out of the hands of the municipality. It was very important that the GSL could act as an independent, impartial party. This was necessary to realise the procedure in which participation was voluntary.

There were some 70 landowners in the area of Hausen-Fasanengarten. Practically all of them wanted to keep their land, i.e. to receive a building plot after the land development. Landowners who wanted to sell their land often did so before the readjustment and the servicing of the area, selling it as *Rohbauland* ('raw' building land). The first landowners played an important role as members of the *Erschließungsgemeinschaft* (land development society). The role of this organisation was to take care of the land development. The municipality too took part in this organisation, because it also owned land in the area. In matters concerning the use of this land, it acted as a private body.

Bois-Guillaume Portes de la Forêt

Two actors played a central role in the development of Portes de la Forêt, the municipality of Bois-Guillaume and a private land developer. Other actors were involved in the process. These were the first landowners, who sold their land at an early stage to the municipality, the private house builders, who realised the largest part of the houses in the area, and the housing corporations, who realised some social housing.

The initiative for the development of the area came from the municipality, who was at that time already the owner of most of the land. This was the result of its policy of land acquisition, with the aim of being able to control its own expansion. The municipality decided to appeal to a private land developer for help. Since the municipality of Bois-Guillaume was only small, it considered that it did not have the necessary knowledge and know-how to realise this kind of operation. Moreover, by choosing a private subdivider as a partner in the process, the municipality would run only very limited financial risks if the commercial development of the area should fail. The procedure of the ZAC allowed the municipality to keep the role of supervisor in the process.

144

Table 4.8   Activities, roles, and actors in Bois-Guillaume Portes de la Forêt

activities ⇒ ⇓ roles	identification of opportunities	land assembly	land development	housing production	ownership/ use
owner	- agriculturalists	- n.a.	- subdivider	- house builders - housing corporations	- owner-occupiers - housing corporations
initiator	- municipality	- municipality	- municipality - subdivider	- house builders - housing corporations	- n.a.
producer	- n.a.	- municipality	- subdivider - sub-contractors	- house builders - housing corporations - sub-contractors	- n.a.
designer	- n.a.	- n.a.	- subdivider -municipality	- house builders - housing corporations - subdivider - municipality	- n.a.

The main activity of the private developer was to carry out the land development. His central aim was to ensure his own continuity and expansion by making a profit from this activity. But this did not mean that he paid no attention to the quality of the residential environment. He took pride in doing his work in a good way. Moreover, this was the best way to ensure new assignments in the future. In the case of Portes de la Forêt, the consultation between municipality and developer was especially important. The municipality was the owner of the land. It was possible for it to transfer that land at a low price to a developer, but that meant that conditions could be attached to the transfer. The private land developer was interested in having the cheap land, but realised he would not get it unless he paid the municipality back, in terms of residential quality.

The first landowners did not play an active role in the development of Portes de la Forêt. But the fact that they sold their land at an early stage to the municipality was very important for the course of the process. Because of this, the price that was paid for the land was relatively low.

The last parties who were involved in the development process were the house builders. First, there were the people who built their own houses. They built 80 detached houses. Then there were the private housebuilding companies. They built the 'grouped' housing, complexes of three to 30 houses, sometimes detached or semi-detached, sometimes terraced and even in apartment blocks. Finally there were the *sociétés de HLM* (housing corporations), who built the social housing. In all, besides the people building their own houses, there were about 15 other house builders on the site. Their influence was limited. They bought building plots, which came together with a set of prescriptions as to building height and density, and design guidelines. These conditions reduced the influence of the house builders on the final appearance of the area.

The main actors in the development process of the Poterie were the municipality of Rennes and the *Société d'Economie Mixte pour l'Aménagement et l'Equipement de la Bretagne* (SEMAEB). In addition, the first landowners and the buyers of the building plots can be distinguished as actors who played a role in the process, but their influence on the course of the process and the outcomes was less important.

Figure 4.3   'Grouped' housing in Bois-Guillaume Portes de la Forêt Rennes la Poterie

146

Table 4.9   Activities, roles, and actors in Rennes la Poterie

activities ⇒ ⇓ roles	identification of opportunities	land assembly	land development	housing production	ownership/ use
owner	- agriculturalists	- n.a.	- SEMAEB	- house builders - housing corporations	- owner-occupiers - housing corporations
initiator	- municipality	- municipality	- municipality	- house builders - housing corporations	- n.a.
producer	- n.a.	- municipality - SEMAEB	- SEMAEB - sub-contractors	- house builders - sub-contractors	- n.a.
designer	- n.a.	- n.a.	- municipality - SEMAEB	- house builders - housing corporations - SEMAEB - municipality	- n.a.

The municipality initiated the development and took care of the land assembly, buying the land directly from the first owners. Then it delegated the powers for the realisation of the housing scheme to the SEMAEB. As a result, both the planning and the actual development work were largely carried out by the SEMAEB, although always in close cooperation with the municipality. The SEMAEB sold serviced plots, the building was mainly done by house building companies and housing corporations, some small parts by people building their own houses. The SEMAEB did not carry out the building work itself. Its role was to coordinate and to finance the operation, within the conditions fixed by the municipality.

The municipality of Rennes pursued an active policy of land acquisition during the decades preceding the development of the Poterie. This happened mostly amicably, sometimes using the pre-emption right given to the city by the procedure of the *Zone d'Aménagement Différé*, and sometimes in the last instance using expropriation rights. The city acquired during this period almost all the land within the municipal boundaries. The aim of this policy was to control the development of the city.

147

The municipality chose not to carry out the development itself for several reasons. Because of the size of its administration, it would have had to engage new employees, for whom there would not necessarily have been any work once the development project was finished. In addition, the statute of the SEM offered more flexibility, allowing it to act more like a commercial actor and enabled it to be more efficient in the interactions with private actors. A final reason why the municipality chose to work together with the SEMAEB was to benefit from its knowledge.

Once the development process was finished, the serviced building plots were mainly sold to private house building companies and to housing corporations (a few plots were sold to people building their own houses, but this number is negligible). These had some freedom regarding the design, but the type and number of houses they were to build were largely determined by the SEMAEB, in cooperation with the municipality. Their influence on the residential environment was therefore limited, their freedom to pursue their own objectives and thus to influence the housing scheme was only small.

4.3   Unravelling the housing development process

Each of the actors directly involved in the process of housing development has his own objectives. These objectives are closely related to both the nature of the actor (public, private, hybrid), and to his role in the development process. In turn, the objectives influence the strategy that the actors follow, and the way in which they carry out the activities. This section describes how the division of roles and activities attributed to the different actors can be analysed, and in which way this influences – through the objectives of the different actors – the strategies that are employed.

Throughout the cases, we have encountered a number of actors who were directly and actively involved in the housing development process. First landowners, local planning authorities, private and public-private land developers, private house builders, housing corporations, higher tiers of government. Each of them has his own interests and objectives, according to which he derives his strategy in the process. Here, we try to characterise the different groups of actors. As explained in chapter two, we are aware of the risks of simplification that such a categorisation implies. However, throughout the cases some characteristics of the different actors recur. It is these characteristics that are emphasised in this section.

First landowner

Because this study focuses on greenfield development, the land was in all cases in agricultural use. Often, the land was owned by farmers. Only in Cramlington North-East Sector were parts of the land already in the hands of house builders at the start of the process. In a similar way, the municipalities in Arnhem Rijkerswoerd, Rennes la Poterie, and Bois-Guillaume Portes de la Forêt owned large parts of the land at the start of the process. Agricultural landowners had no real interest in the development process. After they had sold their land, they no longer took part in the development. They could influence the process only by the price they asked for their land. But the price they received depended for the greater part either on the interaction between supply and demand in the market, or on the expropriation regulations, and not on what the landowners wanted to receive. Most often, their only option was to 'go with the flow', although they did probably got less money if they did not negotiate hard.

Only in the cases where the first landowners choose to remain landowner throughout the process, that is in Bonn Ippendorf and in Stuttgart Hausen-Fasanengarten, did they have more options. Especially in Bonn Ippendorf, the *Umlegung* was imposed on the landowners, so they had no option but to take part in it. But this was – generally speaking – not much of a problem. They knew that in the end they would get a plot back, and the value of that plot would be higher than that of the plot they first had. Of course some landowners were attached to their land, and did not like to exchange it, but the procedure appears to have had a much bigger support from landowners than an expropriation procedure, which would have been the alternative ultimately.

The landowners played a rather passive role in the *Umlegung*: They were subjected to it, without being able to influence it substantially. But once the procedure of the *Umlegung* had finished, and the land development was carried out, they shaped the final appearance of the area to a large extent, because they were responsible for constructing the houses. The objectives of the landowners varied. Some of them only wanted to receive the value increase of the land due to land conversion. These owners usually sold their plot before the procedure of the *Umlegung*. Others wanted to build a house on their new plot. These owners were eager to receive a plot that was well situated. Still others tried to speculate (to the extent that this was possible within the procedure) with their plot. They kept it during the procedure of the *Umlegung*, and hoped that during the procedure the value of the plots would increase (the values used for the calculations in the *Umlegung* were those at the beginning of the procedure, this was fixed in the regulations). After the procedure, these owners sold their serviced plot to others who wanted to build houses. Of course, they were keen on getting a building plot in a nice location, because these plots were likely to yield a higher price. All three types of landowners were to be found both in Bonn Ippendorf and in Stuttgart Hausen-Fasanengarten.

Local planning authority

In all cases, the local planning authorities were responsible for the development plan and the development control. That meant that they supervised the whole development and decided – according to the limits of their power – what development could take place in the area, and in which way. This meant they had a big responsibility towards the (future) inhabitants for the overall residential quality. Most often, the planning authorities also initiated the housing development. They proposed locations where development could take place. In this way they influenced the residential environment at a very early stage of the development process: Through the choice of the site they chose a certain set of site specific characteristics that influenced the residential environment. This choice could also influence the amount of money that became available in the development process and which actor received this money, since this depended partly on who owned the land at the start of the process.

It was usually the stated objective of the municipality to develop a residential area with as high a quality as possible, within the limits set by financial and other constraints. In the case of Arnhem Rijkerswoerd, for

150

example, the municipality took a direct part in the development process with a *bouwteam* (project team) and an *ontwerpteam* (design team). The project team negotiated per section with the developers about what exactly was to be realised in each section. Apart from representatives of the developer and the architect, representatives of the municipal departments most involved took part in this team: urban design, economy, public works, land department. And there was a representative of the regional energy department, because of the importance of this body for the connection to existing (infrastructure) networks. Under supervision of the *projectleider* (project leader) from the municipal department of city planning, this team took the decisions about the realisation of the project. The *ontwerpteam* operated on a somewhat higher level. Its task was to supervise the overall quality of the development by issuing guidelines for the housing that was to be realised.

Public-private land developer

In one of the cases, i.e. Rennes la Poterie, a public-private party carried out the land development. The statute of this organisation needs some explanation. A *Société d'Economie Mixte* or SEM is a public-private body. Its statute is arranged in a special law from 1983. The SEM is a private corporation but at least half of its capital comes from public shareholders. Often the share of public parties is much larger. This was also the case for the SEMAEB, where around 80% of the capital came from public shareholders. The SEM can be seen as a body that pursues public aims, but with the flexibility of a private body. One of the particularities of the SEM in general, and also of the SEMAEB, is that its principal shareholders are also its principal clients. As a result, making profits is not one of the primary aims of a SEM, although it is allowed to do so. An advantage when a SEM instead of a local planning authority plays the coordinating role in a development process is that it is more flexible. It acts more like a private party and does not have to respect the often long public procedures a municipality has to go through when taking decisions.

In the case studies, two types of private land developers could be found. In Bois-Guillaume Portes de la Forêt, an *aménageur-lotisseur* (subdivider) took care of the land development and sold serviced building plots to house builders. In the cases of Bishop's Cleeve and Cramlington North East Sector, private house builders carried out the land development, built the houses on the plots and sold those to final users (most often owner-occupiers). In both cases, these actors wanted to earn income from the value increase due to the land conversion. However, their respective roles in the development process were different.

For the private house builders, it was the sale of houses that allowed them to capture a value increase in the land. They acquired and developed the land and then built houses on it. They earned their income only by selling the houses. It was therefore important to them to follow the trends in the housing market as closely as possible. They tried to build according to demand. This was clearly visible in the cases of Cramlington and Bishop's Cleeve. The house builders made an estimate of how many houses they would be able to sell in a certain period, and they tried to build exactly this number. As a result, they developed this type of large site in sections of 50 to 70 houses. This affected the appearance of the housing scheme. The sections could be identified in the final development as units with a single access road, linked to similar other units only by the main through road. There was no 'overall picture' of what the scheme should look like.

The residential environment was important to the house builders in so far as it influenced people's perception of an area, which would be reflected in the house prices. They tried to build what they thought the market wanted, so that they could sell. Thus they influenced the appearance of the residential environment by their choice of housetypes. They used standard types of houses and fitted them into each new area that was developed. They changed some details here and there, but this still made the areas quite uniform. This had to do with design and was a significant aspect of the residential environment in Cramlington. This observation incited the local planning authority to allocate 'special design sites', sites where something different had to be realised. When the developers came across the special design sites, this posed a problem, because they could not use their standard house types there, they were supposed to use something different. That meant uncertainty about the costs of the houses and about whether or not people would be interested in buying them. Therefore, the house builders were somewhat reluctant to build on those special design sites.

Eventually, the house builders found a way of dealing with these sites that can best be described as a compromise between certainty for the house builder and special design for the planning authority. It is another illustration of the way in which the house builders' dependence on the market influences their strategies and therefore the residential environment. What they did was to build house types from another development they were building in the south of England. This was a good solution for the house builders because they knew exactly what the construction costs of these houses would be. It introduced house types that differed from what was generally built in Cramlington. The house builders then put them in an attractive setting, using garden walls and greenery, and it was a special design site. And, as a representative of the house builders confirmed: 'It works very well, it sells very well'.

In the case of Bois-Guillaume Portes de la Forêt, the private land developer did not build houses, but sold building plots. To ensure his continuity and his expansion, this private land developer had to earn an income from this activity. The developer gave three reasons why he was interested in the quality of the residential environment he realised. First, there was the conviction that doing good work was the best way to ensure new assignments in the future. Furthermore, the company liked to see itself as a craftsman, whose craft is preparing areas for housing development. As such, it took pride in realising a certain quality of residential environment. This culture of quality was equally used as a marketing strategy. Finally, there was the consultation with the municipality. In the case of Portes de la Forêt, this was especially important. The municipality was the owner of the land in the development area. It had the possibility to transfer this land at a low price to a developer, but that meant that it could attach conditions to the transfer. The land developer was interested in having the cheap land, but realised he would not get it unless he paid the municipality back, in terms of residential quality.

Private house builder

In all cases, private house builders realised at least part of the houses. Except for the cases of Bishop's Cleeve and Cramlington, these private house builders bought serviced building plots. They did not directly participate in the process of land conversion. They primarily wanted to build houses and earn an income from that activity. However, they often had some influence on the process of land conversion, for example on the parcellation.

The case of Arnhem Rijkerswoerd illustrates this. To be able to build the houses, the developers had to act to some extent according to the demands of the municipality, because the latter decided which developer got the right to build on a site. The requirements of the municipality had to be realistic, however. If not, no developer would have been interested to carry out the plans. The developers negotiated with the municipality about the norms. Because the house builders' interest was to build houses that could be sold, and the revenues of the house builders consisted of the prices of the houses, they were concerned with the residential environment. What they did was to screen the plans of the municipality to see whether they were realistic in market terms. If not, they expressed their concern to the municipality. But the municipality remained the actor who made the final decision about, for example, location and house type. The only choice for the house builder in the extreme case was either to withdraw or to continue. But this meant that the municipality could not make demands that were unrealistic from a 'marketing' point of view.

In all cases, the main influence of the house builders on the residential environment was by their choice of the design of the houses. This was important for them, because the right choice could mean that the houses could be sold more easily and that higher prices could be asked. This was a concern for the house builders, who made their living from selling the houses. Sometimes it meant that requirements of the municipality with regard to design were partly put aside when the houses did not sell well. It also happened that the houses sold better than expected, which meant that the house builders had fixed a price that was too low.

Housing corporation

In some cases, housing corporations were active as house builders. Their interest in the residential environment was even stronger than that of the private house builders. They were concerned not only with finding a first occupant for the house. The dwellings they built remained their property, so they were also concerned with long term maintenance, and with the question of whether they would still be able to find occupants for the dwellings in 50 years. Their interest in the long term viability of the housing scheme made the housing corporations more concerned with the residential environment than developers of houses for sale. For that reason, they closely followed and tried to influence decisions about the residential environment taken in the development process. They had influence because the planning authorities wanted social housing to be constructed and need the housing corporations for that purpose.

Higher tiers of government

Higher tiers of government usually play a role through the planning system of a country. It is their task to check whether the local plans fit into a broader idea of the development of the region or even the country. Moreover, they often issue guidelines, or minimum standards, for the development based on social or safety considerations. According to the planning system of a country, the way in which these functions are shared by the different tiers of supra-local government varies.

For example, in the cases of Bonn Ippendorf and Stuttgart Hausen-Fasanengarten, the role of the *Bundesregierung* (federal government) remained only in the background. There were no overall spatial plans on the federal level, with which the lower level plans had to comply. But federal government offered guidelines for development that had to be taken into account by lower levels of government in the *Baunutzungsverordnung* (Federal Land Utilisation Ordinance). This *Baunutzungsverordnung* prescribed what it was possible to arrange in (local) land use plans. Although it left a substantial margin for the lower levels of government to follow their own policy, it determined to some extent what could and what could not be realised. It did this by specifying the different types of land uses that the local plan could prescribe. For each type, e.g. residential only areas, general residential areas, mixed use areas, commercial areas, it prescribed what types of land use were permitted. The *Baunutzungsverordnung* offered the tools with which the local land use plan was drawn up. The reason why federal government did this was to provide the whole of Germany with a uniform interpretation of the various urban land use plans. In addition to this, for social and safety considerations, central government issued minimum standards for the construction of roads and buildings in the *Musterbauordnung* (Model Building Code). These were only minimum norms, and they left a margin for the municipalities to pursue their own policy.

The *Landesregierung* influenced the housing development process only indirectly, through the local land use plan. The *Landesplanung*, a spatial development plan issued by the *Land*, for the whole of its territory, gave broad guidelines for the development of the city. It had to be taken into account in the *Flächennutzungsplan*, and hence in the local land use plan. According to the German planning law, higher level plans were binding for lower level plans, within the administration. Only the local land use plan was legally binding for citizens. This meant that the *Landesplanung* had to be taken into account by the municipalities, but because of its broad, strategic character, its influence on the actual development process was small.

In other cases, the way in which these aspects were arranged differed. But the responsibility of higher tiers of government for the embedding of the development on the local level in a general plan, and the social and safety standards, could be found in every case. This responsibility usually worked through the local plan, which had to take into account central or regional government plans and/or guidelines.

In the case of Arnhem Rijkerswoerd, central government played a much more direct role: Without the *locatiesubsidie* from central government it would not have been possible to realise Rijkerswoerd, certainly not in the final form. Before granting this subsidy, central government imposed minimum requirements concerning the layout of the development, the residential environment was also influenced by central government.

Central government would subsidise only when it was certain that its money was used effectively. To control this it set norms. Basically, the municipality and representatives of central government agreed upon the number of houses to be built in Rijkerswoerd, and upon a division of social and market sector houses. To calculate the costs that the municipality was to incur, prices were agreed for each activity. For example the construction of a road would be calculated at a certain cost per metre, each street light had a certain cost, etc. The revenues from the sale of building plots were calculated. For the difference between these revenues and the costs of developing the area, the municipality applied to central government for *locatiesubsidie*. Central government would then grant this. But if it appeared later that the actual housing scheme the municipality realised did not comply with what had been agreed upon, the money would have been claimed back. So the fact that the municipality applied for *locatiesubsidie* limited its possibilities to influence the residential environment.

4.4 Power and dependence relations

As stated in chapter one, this study takes the point of view that no single person is able to control decision making in an 'interactive policy network'. In this study the housing development process is considered as such a network, in which the different actors need the others to realise their objectives. There is a mutual dependence. In this section, we elaborate the way in which our analysis deals with the mutual dependence. Elias (1971) describes dependence as a fundamental aspect of human interaction, and dependence relations between persons as the inverse of power relations. Whenever two or more persons interact, there is a certain 'power balance' that determines the possibilities of each of the persons to influence actions of the others. This notion of power and the way in which it influences the behaviour of the actors in a situation of mutual dependence is central to our understanding of the decision making in the housing development process. The literature about power is vast, and there is no single definition of what power is. For that reason we develop first the way in which this study deals with power in general terms. Then we investigate how this can be analysed in housing development processes.

For our analysis of power, we refer mainly to Giddens (1984). He stresses that power plays a role in all human actions when he says that: 'Action depends on the capability of the individual to "make a difference" to a pre-existing state of affairs or course of events. An agent ceases to be such if he or she loses the capability to "make a difference", that is, to exercise some sort of power' (Ibid.: 14). Saying that no single actor is in charge in a housing development process, and that all the actors depend to varying degrees upon the others is at the same time saying that all the actors have some resources by which they can influence the behaviour of others. Giddens calls this the 'dialectic of control' in social systems.

To develop this view of power, Giddens (in Cassell, 1993) refers to Parsons' later work on power. Parsons criticises a 'zero-sum' concept of power, which conceives power as '... to be possessed by one person or group to the degree that it is not possessed by a second person or group over whom the power is wielded' (Giddens in Cassell, 1993: 212). Parsons proposes another view of power, i.e. power can be generated in a social system. He compares it to wealth being generated in the productive organisation of an economy. He stresses that possession and use of power can be expressed in interactions through negative sanctions (punishment, or the threat of punishment) and positive sanctions (offering something the other wants). But sanctions are not necessarily employed when power is used in the interactions. Notably the negative sanctions are largely symbolic: in stable political systems they are only employed as a last resort. Giddens adds some critical notes to Parsons' analytical separation of the amount of power a person or group holds, and the question of what sanctions it are used to enforce power, but that is not so much our concern here. Where we want to get to – and Giddens agrees on this – is that Parsons work underlines that '... the use of power frequently represents a facility for the achievement of objectives which both sides in a power relation desire' (Giddens in Cassell, 1993: 219).

Giddens sustains Parsons' critique on a 'zero-sum' conception of power. But he does not follow Parsons' reconstruction of the notion when he conceptualises power in his structuration theory. Giddens stresses that 'Power is not intrinsically connected to the achievement of sectional

158

interests' (Giddens, 1984: 15), and that in this view '... power is not itself a resource' (Ibid.: 16). He concludes that '... all forms of dependence offer some resources whereby those who are subordinate can influence the activities of their superiors' (Ibid.: 16). These resources can be divided in allocative and authoritative resources. The first type of resources corresponds to the traditional conception of resources as it refers to material objects – either raw materials or produced goods – and means of material production. Authoritative resources are more complicated to conceptualise. It refers to the rules that guide actors in mutual association or in self-development and self-expression (Ibid.: 258). In this study, Giddens' conception of power is used as a starting point for the analysis of decision making about expenditure on the residential environment during the housing development process.

Power relations in the housing development process

Once the view of power developed above has been accepted, the next step is to find a way to analyse it. In an attempt to apply Giddens' ideas on structure and agency to the analysis of the development process, Healey (1992b) speaks of resources and rules as the equivalent of Giddens' division into allocative resources (resources) and authoritative resources (rules). Resources and rules constitute the link between structure and agency. Healey translates the notions of structure and agency to the development process. In her view, agency refers to the activities of the actors in the development process. Structure then concerns the context in which these activities are employed: capital accumulation and regulation. In these terms, resources and rules constitute the link between the activities of the actors and the structural context in which these take place. Healey then introduces a third link that we assent to, which is constituted by the '... ideas used in defining and developing a project within the context of the prevailing rules and resources' (Ibid.: 35). There is a certain parallel between this division of bases for power into resources, rules, and ideas and the often used classification of ways in which actors can influence the behaviour of others in policy processes, that is by economic, juridical, and communicative control (see Verhage and Needham, 1997). We use this similarity to translate the above notions into analytical tools that allow us

159

to distinguish the power and dependence relations between the actors in the housing development process.

Resources in housing development are for example land rights, labour, finance, and information. Actors who possess these resources can direct – to a certain extent – the actions of the others: They have (for example) the money the other actors need to realise their objectives, or they have crucial information, or land. An actor who expresses his power in the housing development process through these resources can be said to have economic power. Conversely, actors can be said to be economically dependent upon the actors who command the resources they do not have.

Rules govern the way material resources are used. These rules may be formalised in laws or administrative procedures, but they can also exist only in custom or practice. To explain this difference, Ostrom (1986) uses the distinction between formal rules and working rules. Formal rules are an important source for working rules, but working rules can also be based on cultural or social ways of behaving that are found to be appropriate. In that case, following them can be a 'social habit' (Ibid., 1986: 466). That means that the rules that govern behaviour of actors are broader than only the legal regulation. Rules can be a basis for power also when they are not juridically codified. For practical reasons, however, our analysis concentrates on the rules that are codified, and thus form a basis for juridical control or power and dependence. In Ostrom's terms, these are the formalised rules. More implicit working rules, e.g. as to how people should work together, or what the role of each actor should be, what type of interaction is considered appropriate in a particular situation, are described only when they offer a clear or an additional explanation why things went as they did in the studied cases. Bringing to light all the working rules that actors follow goes beyond the scope of this comparative study, and would require a deeper, more cultural analysis.

Formal rules, or in our terms juridical power or dependence, can be divided into public law – that enables a public body to impose actions on others – and private law – that regulates obligations freely entered into. It may be clear that rules deriving from public law regulate parts of the housing development process. In the studied cases, there is always a public authority involved in the development process, even if it were only for the granting of a building permit. But the case studies show that actors in the

160

process often 'choose' to work together, that is, they are not 'forced' to do so by public law. Since private law regulates obligations that parties freely enter into, rules that derive from private law also play an important role in the housing development process, both in the relationship private/private and in the relationship private/public.

Ideas can be an important basis for power or control in the development process. Healey (1992b: 35) formulates it as follows: '...ideas (...) inform the interests and strategies of actors as they define projects, consider relationships and develop and interpret rules'. This means that an idea of one of the actors can also influence the behaviour of the others. Actors who have an idea with a great binding force in the development process thus control – on the basis of this idea – the others. In other words, they have power. This comes close to what is sometimes called communicative control, actors influencing the behaviour of others by providing or withholding information, or by the force of ideas. This is the third basis for power with which this study is concerned. The case studies exemplify it further.

The above description, combined with Elias' (1971) observation that power and dependence are each other's inverse, can be represented in the schedule in table 4.10. When filled in with empirical data, this framework gives insight into the way in which the actors influence each other.

Table 4.10   Possible power and dependence relations between actors in the development process

Actor 1	Relation	Actor 2
resources (e.g. land rights, information, finance)	A1 depends on A2 for ... A2 depends on A1 for ...	resources (e.g. land rights, information, finance)
rules (e.g. public law, private law)	A1 depends on A2 for ... A2 depends on A1 for ...	rules (e.g. public law, private law)
ideas	A1 depends on A2 for ... A2 depends on A1 for ...	ideas

4.5   Case study files: relations of power

Arnhem Rijkerswoerd

The municipality of Arnhem acquired the land from the first landowners a long time in advance of any development. Because this was common practice, and because the municipality had the possibility to use expropriation, the landowners were in a rather weak position. They did not have much power to claim anything from the municipality, or at least, they did not exercise it.

As a temporary landowner during the land development, the municipality had quite a lot of power over the other actors in the process. As a public body, it had power under public law. But in this case, it also used powers based on its resource of land when it attached conditions concerning the development to the disposal of plots. Because the municipality had the resource of land, the developers had to come to the municipality to acquire the necessary land to build houses.

Table 4.11   Relations between the landowners and the municipality

Landowners	Landowners depend on municipality	Municipality depends on landowners	Municipality
Resources - land	Only if they are willing to sell their land (municipality had a "monopoly of demand")	If they decide not to sell for the price that is offered, this obstructs the municipal policy	Resources - money
Rules - right of ownership	If they do not cooperate, they can be expropriated	If the landowners do not cooperate, this leads to time consuming and costly expropriation procedure	Rules - expropriation
Ideas - n.a.	n.a.	The policy can only be pursued through the landowners	Ideas - active policy of land acquisition

This also worked the other way round. The municipality owned the land, but could not build houses. It needed developers (either private or housing corporation) to realise its objectives. If the municipality's requirements were unrealistic (in terms of market conditions, or technically), the latter could decide not to cooperate. This position was even stronger for the corporations than for the private house builders, since the latter could more easily be replaced by another developer, as there was more competition.

The partners of the municipality in the development process had some power over the municipality. The latter acquired the land, for which it applied to banks for finance. So the municipality had to pay interest charges. The longer it took between the acquisition of the land and the disposal of the plots, the higher the interest costs would be. If the municipality was engaged in the development of a part of the plan with a developer it was possible that after a certain period it appeared that the parties could not reach an agreement. But if the municipality were to start all over again with a new partner, it would loose a lot of time, hence money, in the form of interest charges. So just by being invited to develop part of the plan, the developer acquired some negotiating power.

Table 4.12 Relations between the municipality and the house builders

Municipality	Municipality depends on house builders	House builders depend on municipality	House builders
Resources - land	To realise the plans, house builders are needed to build houses	Without the land, house builders cannot proceed	Resources - money - know-how
Rules - local plan - conditions of sale	n.a.	House builders must act in accordance with local plan to get building permission. More detailed requirements can be attached to the sale of the land	Rules - n.a.
Ideas - image of residential quality	The house builders' knowledge allows realising houses according to demand	The municipality's helicopter view allows to get coherence in the housing scheme	Ideas - attracting buyers/tenants

Zwolle Oldenelerbroek

The municipality had relations with other parties, the private house builders, the housing corporations and the first landowners. The first landowners owned the land, the central resource in the development process. This made the municipality depend on them. The landowners used this dependence to claim a higher price for the land than the value in its existing use. However, their power was severely restricted by 'rules'. The municipality could use expropriation to obtain the land. It did not do this, but this functioned as an incentive for the landowners to sell their land quickly at a price that was judged reasonable by both parties.

Both the private developers and the housing corporations had little margin to influence the housing scheme. Their task was to realise the housing scheme proposed by the municipality. They did have a certain margin of freedom as to the architectural aspects of the houses they built.

But before granting a building permission, the municipality could check whether the proposed architecture could be accepted in the scheme. An independent committee was created (the *Welstandscommissie*) especially to judge the architectural quality of the proposed plans.

Described in this way, the working relations between municipality and developers sound more hierarchical than they were in practice. For each section of the plan, municipality, developers, and housing corporations negotiated about the layout and form as a project team. In this team, each of the participants was represented, the developers by a project leader the municipality by a representative of each of the departments concerned. Within this team, the meetings of the different representatives were coordinated by the municipal project leader.

Represented schematically, the relations of dependence between the actors in Zwolle Oldenelerbroek were identical to those in Arnhem Rijkerswoerd, see tables 4.11 and 4.12.

Bishop's Cleeve

When the development of the area was first proposed, quite a lot of land was still in the hands of the first landowners. This gave them economic power over the house builders, who depended on the first landowners to get the land they wanted for the development. The first landowners used this power to obtain a price for their land that was quite a lot higher than its agricultural value.

Once the house builders owned the land they were in a position to decide relatively freely when they were going to build and what exactly they were going to build. Of course, this power was limited by the planning authorities who used their statutory power (under public law) to control the development. In addition, the planning authorities − mainly the Borough Council in its role as controller of the development − used communicative power to try and influence the house builders. This power consisted of trying by persuasion to influence the actions of the house builders.

The County Council − as the highway department − did not often use persuasion to achieve its objectives. Its powers over the design and construction of the roads were more heavily backed by statutory powers. The highway network is a much more clearly defined object than the residential environment, and its requirements are correspondingly clearer: Road safety stands 'above discussion'.

Table 4.13   Relations between the house builders and the first landowners

Landowners	Landowners depend on house builders	House builders depend on landowners	House builders
Resources - land	If they are willing to sell, landowners depend on house builders to get a good price	Land is necessary for building houses. Land use planning restricts possibility of substitution	Resources - money
Rules - rights of ownership	n.a.	If landowners do not want to cooperate, house builders have no means to obtain the land	Rules n.a.
Ideas n.a.	n.a.	n.a.	Ideas n.a.

Besides their power under public law and the communicative power, both the County Council and the District Council used private law, they made agreements with the house builders. However, this possibility of drawing up agreements with the house builders was created by a regulation in public law. House builders entered into these agreements because they were linked to the planning permissions: If the house builders had not agreed to enter into a private agreement, the Borough Council could have refused planning permission. On the other hand, the agreements offered the house builders security about what they had to provide, which was very important as a basis for their calculations.

Table 4.14 Relations between the local planning authorities and the house builders

Local Planning Authorities (LPA)	LPA depend on house builders	House builders depend on LPA	House builders
Resources - n.a.	House builders are needed to realise a housing scheme (but they were eager to develop here)	n.a.	Resources - land - building capacity
Rules - development control - planning agreements	n.a.	Land needs to be designated for housing. To obtain permission, agreements have to be adhered to	Rules n.a.
Ideas - image of residential quality	House builders must be able to bring into agreement LPA's image of residential quality with demand for housing	LPA is not able to impose its image of residential quality other than by persuasion	Ideas - attracting buyers

Cramlington North-East Sector

The house builders were in a comfortable position because they owned the land at the beginning of the development process. They had acquired it for what they call a 'historical price', a price that was not much higher than the agricultural value of the land. Because they had this central resource, the house builders had quite a lot of power. The planning authority depended on them for the realisation of houses. On the other hand, the developers had to build houses for the continuity of their business. So there was a sort of equilibrium in which house builders and the planning authority decided up to which 'quality level' the housing scheme should be realised.

This equilibrium worked for two reasons. Firstly, because the house builders had acquired the land for a low price. The financial margin was high, so it was not too much of a problem for them to spend some of it on, for example, a cycleway network or on public parks. Secondly, because

Cramlington was developed as a new town, there were hardly any facilities in the area before the development started. The house builders knew that and they realised from the beginning that if they were to develop this area, they had to provide the necessary facilities.

The planning authority had its powers from its statutory function as the producer of the local plan and as the body that is responsible for the development control. The final outcome of the development process was thus mainly a product of two interdependent parties interacting. In general, both parties agreed that the relations have generally been good during the development process, which again illustrates the equilibrium that the parties seem to have found. A reason for the generally good relations was the fact that there was some 'slack' – as a house builder put it – because of the low price for which the house builders had acquired the land.

Represented schematically, the relations of dependence between the actors in Cramlington North-East Sector were identical to those in Bishop's Cleeve, see tables 4.13 and 4.14.

Figure 4.4   **A children's play area in Cramlington North-East Sector**

The instrument of the *Umlegung* allowed the municipality – although it did not own the land – to have a very direct influence on the realisation of its own plan. The *Umlegung* did not imply any compulsory purchase. That meant that the landowners continued to play a role throughout the development process, but their powers were restricted. They could not withhold their land from the development process, because the procedure of the *Umlegung* gave the municipality legal powers to force them to participate. This removed part of the power based on their main resource, the land.

The municipality was the biggest single landowner in the area. Although it did not follow a land banking strategy, it did acquire land throughout its territory. Sometimes actively, but mostly passively: When a landowner offered a piece of land for sale, and the municipality found it fitted into its long term development strategy (as presented in the *Flächennützungsplan*), it bought the land. The fact that the city of Bonn owned these plots did not increase its power over other landowners. It did, however, facilitate the task of the municipality to make a new parcellation. These plots allowed the municipality to propose different parcellations to the landowners. Because it owned some land, the municipality had the possibility to exchange plots with owners of land in less favoured locations. This made it easier to carry through the *Umlegung* because when the landowners received a plot out of the *Umlegung* that satisfied them, they were not likely to use their procedural possibilities to ask for reconsideration of the parcellation proposed by the municipality. In that sense, the ownership of land gave the municipality more influence in the procedure.

The landowners were often also the house builders in Ippendorf. Exceptionally, building constructors bought several plots to construct apartments. However, no single actor acquired more than four plots. Therefore, there was hardly any influence of the house builders that could be distinguished separately from that of the landowners. The house builders/landowners did influence the final appearance of the area, because of the design of the houses they chose. The municipality of Bonn left them rather free in this choice. Few prescriptions as to the design of the houses were included in the local plan. Also, no plots were indicated as plots for social housing.

Figure 4.5   An appartment building with several dwellings on one plot in Bonn Ippendorf

Table 4.15   Relations between the landowners and the municipality of Bonn

Landowners	Landowners depend on municipality	Municipality depends on landowners	Municipality
Resources - land	n.a.	Municipality needs to reparcel the land to realise its land use/housing policy	Resources - n.a.
Rules - rights in procedure of *Umlegung*	Municipality draws up new parcellation and determines future land use. To obtain planning permission, landowners have to comply with demands of the municipality	The new reparcellation must do justice to the claims of the landowners. If not, the latter can obstruct the procedure	Rules - procedure of *Umlegung*
Ideas - n.a.	n.a.	n.a.	Ideas - n.a.

Stuttgart Hausen-Fasanengarten

The particularity in this development process was that the municipality kept the landowners dependent by withholding the change of land-use in the area. In fact, this gave the municipality economic power: It was up to the municipality whether or not a value increase of the land would take place, from which the landowners could benefit. There was a mutual dependence between the municipality and the landowners in this case. Without the cooperation of the landowners, the process could not have been carried through. This meant that the municipality and the landowners had to enter into negotiations to find a solution that was acceptable to both parties. When signing the agreement, the landowners accepted the development plan proposed by the municipality. This agreement was necessary to ensure the continuation of the development process.

As an independent party, the appointed private intermediate in the process, the *Gesellschaft für Stadt und Landesplanung* (GSL) did not have any authoritative or allocative power (rules and resources). However, its role was crucial for the success of the operation. It was the task of the GSL to transmit the municipality's ideas to the landowners, and to convince them that participating was advantageous. The way this worked out is a good illustration of the binding force of ideas in the development process.

Table 4.16   Relations between the municipality of Stuttgart and the first landowners

Landowners	Landowners depend on municipality	Municipality depends on landowners	Municipality
Resources - land	Municipality decides whether value increase of land takes place	Municipality needs the land to realise its land use/housing policy	Resources - potential value increase of land
Rules - right of ownership	In the phases of land development and housing construction, landowners have to comply with municipal requirements to obtain planning permission	Landowners have to agree upon the proposed conditions to bring in their land	Rules - voluntarily agreed conditions - development control
Ideas - capturing the value increase of the land	If municipality withdraws from the process, landowners do not gain anything empty hands	Value increase that remains for the landowners needs to be high enough to make them participate	Ideas - inventive use of public powers to create negotiating position

Once the agreement was signed, the actual work for the reparcelling and the servicing of the area could start. The landowners were themselves responsible for this work. In this phase, when the *Umlegung* was finished, and the area was ready to be built upon, the task of the GSL was finished. The municipality then entered into a process of interactions with the house builders. The basis for these interactions was the land use plan. This plan was a political document. Although the preparation was done by the technical departments in the municipality, the final decisions were taken by the municipal councillors.

Table 4.17   Relations between the municipality of Stuttgart and the GSL

Municipality	Municipality depends on GSL	GSL depends on municipality	GSL
Resources - n.a.	n.a.	n.a.	Resources - n.a.
Rules - n.a.	n.a.	n.a.	Rules - n.a.
Ideas - recouping part of the value increase in the land to use for primary and secondary services	GSL must bring the ideas of the municipality across to the landowners	Municipality must have realistic demands, which are a reasonable bid towards the landowners	Ideas - knowledge of how to approach and convince landowners

Table 4.18   Relations between the landowners and the GSL

GSL	GSL depends on landowners	Landowners depend on GSL	Landowners
Resources - n.a.	n.a.	n.a.	Resources - n.a.
Rules - n.a.	n.a.	n.a.	Rules - n.a.
Ideas - knowledge of how to approach and convince the municipality	Landowners must have realistic demands, so that the municipality carries through the *Umlegung*	GSL must bring the ideas of the landowners across to the municipality	Ideas - capturing the value increase of the land

Figure 4.6   An appartment building with social housing in Stuttgart Hausen-Fasanengarten

Bois-Guillaume Portes de la Forêt

The municipality bought the land before the development process actually started. To do so it partly benefited from incidental offers from landowners to sell, but it also declared the land as a 'public utility', which enabled it to use (public) powers of expropriation. Although this was not actually used, this played a role in the power balance between municipality and first landowners. It allowed the municipality to buy the necessary land at a reasonable price. Because the municipality owned the land at the start of the process, it had a strong position in the development process: It could decide which private partner to choose, and it could impose conditions on the private partner in exchange for a moderated disposal price for the required land.

Table 4.19   Relations between the landowners and the municipality of Bois-Guillaume

Landowners	Landowners depend on municipality	Municipality depends on landowners	Municipality
Resources - land	Only if they are willing to sell their land (municipality had a "monopoly of demand")	When landowners decide not to sell for the price that is offered, this obstructs the municipal policy	Resources - money
Rules - right of ownership	If they do not cooperate, they can be expropriated	If the landowners do not cooperate, this leads to time consuming and costly expropriation procedures	Rules - pre-emption - expropriation
Ideas - n.a.	n.a.	The policy can only be pursued through the landowners	Ideas - active policy of land acquisition

The developer had the required technical and economic knowledge to realise a housing scheme, knowledge that the municipality did not have. However, in the procedure of the *Zone d'Aménagement Concerté* (ZAC), the municipality followed closely what the developer was doing. In this procedure, all the decisions that had to be made in the development process were made in consultation, or negotiation, between the partners. The task of the developer was to realise a high quality, but to do that in an operation that was commercially viable. The low land price created conditions to do so in Bois-Guillaume. Thus, a power balance that appears to have pleased both emerged between the two main actors in this project. This balance was laid down in the convention signed between the two at an early stage of the process.

The house builders bought the developed land from the land developer. They were not free to do what they wanted with their plots. First, they had to pass the test of the building permission. Besides this development control, there were design principles, to which the house builders had to agree when they bought one or more plots. They were in a position of dependence, and did not have much 'room to play with'.

Table 4.20   Relations between the municipality of Bois-Guillaume and the land developer

Land developer	Land developer depends on municipality	Municipality depends on land developer	Municipality
Resources - money - know-how - development capacity	Land developer needs land to do its work	The land developer has the knowledge and capacity to develop the area.	Resources - land
Rules - private law agreement	Land developer must operate within the margins set by municipality to obtain building permission	Municipality must stick to its obligations (mainly as to time periods) agreed with developer. If not, the latter can withdraw	Rules - ZAC procedure: local plan and supervision - private law agreement
Ideas - inventive technical solutions to realise quality within financial constraints	Municipality must be open to proposed technical solutions (possibly deviate from pre-fixed standards)	Developer must find a solution to realise quality within financial constraints	Ideas - image of residential quality

Table 4.21   Relations between the land developer and the house builders

Land developer	Land developer depends on house builders	House builders depend on land developer	House builders
Resources - building plots	to dispose of the building plots	to obtain a building plot	Resources - money
Rules - design guidelines backed by local plan - conditions fixed to sale of plots	n.a.	House builders must comply with design guidelines to obtain land and building permission	Rules - n.a.
Ideas - attracting buyers	House builders must want to buy the plots	Land developer determines parcellation of the housing scheme which influences possible housing types and form	Ideas - realising housing that corresponds to their wishes

Rennes la Poterie

During the land assembly, the municipality had to reach an agreement with the first landowners about the acquisition of the land. For this purpose, it had at its disposal the pre-emption right under the procedure of the *Zone d'Aménagement Différé*. Of course, the first landowners, all farmers in this case, were not always happy with the expansion of the city. However, the municipal council judged its active policy of land acquisition as being more important and used public powers to purchase the land. Once the landowners had sold their land, they had no further influence in the development process. Represented schematically, the relationship between the municipality and the first landowners was identical to that in Bois-Guillaume Portes de la Forêt, see table 4.19.

Before the actual development started, the municipality transferred the land to the public-private *Société d'économie mixte pour l'aménagement et l'équipement de la Bretagne* (SEMAEB). Thus, the SEMAEB obtained a central role in the development. It was this public-private body that carried out the development and administered the budget. The municipality even delegated some of its legal powers to the SEMAEB, to give it the tools to carry out the development. For this purpose, the municipality and the SEMAEB drew up an agreement, in which some major points concerning the role of both parties and the relationship between them in the development process were fixed. The SEMAEB carried out the planning for the area, and became *maître d'ouvrage*, which meant that all the work was carried out under its supervision. However, there was a close relationship of cooperation and to some extent of supervision between the municipality and the SEMAEB. This relationship was formalised in the procedure of the *Zone d'Aménagement Concerté (ZAC)*.

Figure 4.7   The central, more dense part of Rennes La Poterie

178

The SEMAEB sold the building plots to the house builders. Because of this, it had the possibility of influencing to some degree what the latter built on the plots. These influences on the design of the houses were supplementary to what was already fixed in the regulations attached to the local plan. Represented schematically, the relations between the SEMAEB and the house builders were the same as those between the private developer and the house builders in Bois-Guillaume Portes de la Forêt, see table 4.22.

Table 4.22   Relations between the municipality of Rennes and the SEMAEB

Municipality	Municipality depends on SEMAEB	SEMAEB depends on municipality	SEMAEB
Resources - land	SEMAEB has the technical knowledge and capacity required for the realisation of the housing scheme	SEMAEB needs land to develop, mainly to ensure its subsistence	Resources - know-how - building capacity
Rules - local plan (within ZAC)	n.a.	SEMAEB has to act in accordance with local plan	Rules - n.a.
Ideas - n.a.	n.a.	n.a.	Ideas - n.a.

4.6   Interdependence between the actors

As described in section 4.4, the actors in the development process are interdependent. Each one of them needs the others to realise his objectives. The relations of interdependence are shaped by and in turn shape the development process. In this section, this dialectic of control – to use Giddens' words (1984: 16) – is analysed. In the case study files above, we have seen that the interdependence between the actors does not mean that all the actors are equal. In most cases the interdependence is asymmetrical. There are several reasons for this. In this section we take a closer look at the way in which the different actors in the process can influence the behaviour of the others. Following the classification of 4.4, we focus on resources, rules, and ideas. These categories of power are employed according to the characteristics of both the actor himself, and of the role the actor plays in the process.

Resources

The key resource in the process of housing development is the land. The central role of the owner of the land appears in almost all the cases. Take the example of Rennes la Poterie: Once the first (agricultural) landowners had sold their land to the municipality, they had no further influence on the development process. Before the actual development started, the municipality transferred the land to the SEMAEB. Thus, the SEMAEB obtained a central role in the development.

The relations between the SEMAEB and the municipality concerning the financial aspects of the operation require some further explanation. The accounts of the operation were kept completely separate from the general accounts of the SEMAEB. The income of the SEMAEB consisted of a certain percentage of the revenues from the sale of the building plots, and a certain percentage of the costs of the works that had to be carried out. These percentages were fixed in advance at four percent and three and a half percent respectively. Besides, the SEMAEB calculated a certain price beforehand for the studies it had to undertake in order to realise the housing scheme. This figured as costs on the financial accounts of the operation. There was no permeability between these accounts and the general budget of the SEMAEB: A possible financial margin (or loss) was either received (or borne) by the municipality. For that reason, the municipality had a great interest in closely following the whole operation.

The SEMAEB sold building plots to house builders. This gave it the possibility of influencing to some degree what the house builders built on the plots. This was fixed in the *cahiers de cession des terrains* (document of land disposal), signed at the transfer of the plots. Two articles of this document were concerned with the architectural aspects of the development. It stated for example that the project should be 'integrated in and in coherence with the surrounding plan area'. It was agreed that a planner appointed by the developer checked whether the required coherence was realised. More informally, the SEMAEB influenced the architects that the house builders choose. However, this did not go very far because the SEMAEB realised at the same time that it would not make much sense to force house builders to work with architects they did not want to work with. The ownership of the land gave similar possibilities for such additional requirements in the cases of Arnhem Rijkerswoerd, Zwolle Oldenelerbroek, and Bois-Guillaume Portes de la Forêt.

Other cases illustrate the central role of the resource of land in a different way. For example in Bishop's Cleeve, where the first landowners used their economic power over the house builders to obtain a price for their land that was quite a lot higher than its agricultural value. After this, the first landowners disappeared from the scene and the house builders then acted as landowners. This gave them the opportunity to play a leading role in the development. They could decide to quite a high extent when they were going to build and what exactly they were going to build. Of course, this power was limited by the planning authorities who used their statutory power (under public law) to control the development.

Other types of resources play a role in the housing development process, for instance money. Often, this is used to obtain land and hence a role in the process. For house builders or developers, their capacity to build houses or develop land is a resource which gives them power over local planning authorities that do not posess these resources. In a similar way, specific technical or economic knowledge is a resource that is often used in the process by actors other than the local planning authority.

In Stuttgart Hausen-Fasanengarten, the municipality used as a resource a potential value increase, which it was able to cause by designating an area for housing development. The municipality kept the landowners dependent to some degree by withholding the change of land-use in the area. The municipality made a draft plan, but the procedure to adopt the plan officially was not started until all the landowners had signed an agreement in which their contribution towards the land development and the residential environment was fixed. Only then was the land use plan adopted and the procedure of the *Umlegung* officially initiated. Of course, the municipality and the landowners were mutually dependent. Without the cooperation of the landowners, the process could not have been carried through. This meant that the municipality and the landowners had to enter into negotiations to find a solution that was acceptable for both parties.

Several types of rules that play a role in housing development can be distinguished. The rules employed by the planning authority as a public actor are the most obvious. They allow the planning authority to carry out its function of development control by issuing planning permissions, or they allow it to purchase the land required for development by expropriation or pre-emption. An example of the latter type of rules was used in the development of Rennes la Poterie. In this case, the municipality took care of the land assembly before the actual development started. During this phase, the municipality had to reach an agreement with the first landowners about the acquisition of the land. For this purpose, it had at its disposal the pre-emption right under the procedure of the *Zone d'Aménagement Différé*. Of course, the first landowners, all farmers in this case, were not always happy with the expansion of the city. However, the municipal council judged its active policy of land acquisition as being more important. The municipality's role of accommodating urban growth was considered more important than the retaining of farmers within the municipal boundaries.

Although it was not actually used, the expropriating power of the municipalities also played a role in Bois-Guillaume Portes de la Forêt, Zwolle Oldenelerbroek, and Arnhem Rijkerswoerd. The fact that the municipalities had the possibility of using expropriation influenced the behaviour of the landowners. They knew that eventually the municipality would obtain their land, and this gave the latter a much stronger position in the negotiations about the price at which the land would be purchased 'amicably'.

There was a particularity in the case of Rennes la Poterie in that the municipality delegated some of its public powers to another (public-private) actor, i.e. the SEMAEB. For this purpose, the municipality and the SEMAEB drew up an agreement, in which some major points concerning the role of both parties and the relations between them in the development process were fixed. Here, another type of rule appears, i.e. rules that both actors agree upon and sign, so that these rules become valid under private law. Since the municipality wanted to retain supervision of the development, this delegation of power was surrounded by a number of qualifications. It was agreed that the SEMAEB would carry out the planning for the area, and that it would become *maître d'ouvrage*, which meant that all the work would be carried out under its supervision. However, although it did not administer the budget of the operation, the municipality remained responsible for possible gains or losses. For that reason, there was a close relationship of cooperation and to some extent of supervision between the municipality and the public private developer.

The relationship between the developer and the municipality was formalised in the procedure of the *Zone d'Aménagement Concerté (ZAC)*. This procedure consisted of two main phases, the creation and the realisation. In the first phase, the municipality sought a partner for the development and discussed the way in which the housing scheme was to be developed. In this phase, the municipality took the initiative. The convention between the SEMAEB and the municipality was the end of the phase of creation. In Bois-Guillaume this rule was also used, and a similar convention was drawn up between the municipality and the (private) land developer.

Then followed the phase of realisation. In this phase, the SEMAEB took the initiative. It drew up a *Dossier de Realisation* (realisation file). An important component of this was the budget estimate. Once this had been approved by the municipality, it became the document of reference for the rest of the operation. Each deviation had to be discussed with the municipality. Moreover, a *comité de pilotage* (steering group) met during the process on a two-monthly basis to check the proceedings of the operation.

In other cases too, rules that are valid under private law were used to tie the different actors to an agreed form of the development process and/or the housing scheme. For example in Cramlington North-East Sector and in Bishop's Cleeve, besides their power under public law, the local planning authorities used private law to enforce their powers: They made agreements under private law with the house builders. In these agreements they arranged the ways of financing secondary services, such as school sites, parks, and the bypass. The house builders entered into these agreements because they were linked to the planning permissions: If the house builders did not follow the agreement, the Borough Council could refuse planning permission. This gave the agreements a hybrid character, they functioned as private agreements, but it was public regulation that allowed the planning authorities to involve private parties in such agreements.

Private law agreements were also used in Stuttgart Hausen-Fasanengarten. For example, the landowners had to sign an agreement in which they accepted the development plan proposed by the municipality. Otherwise it would have been possible for them to object to the plan in the official planning procedure, which was only started after the agreement had been reached. Normally in Germany, any legal person can raise objections to a local plan. In the *freiwillige Umlegung* as it was realised in Hausen, such a possibility would have thwarted the agreement reached between the municipality and the landowners. The agreement in which the landowners agreed to accept the plan was necessary to ensure the continuation of the development process. As a result, only third persons could object to the plan, but this did not happen.

The *amtliche Umlegung* which was used in Bonn Ippendorf can be seen as an alternative to the possibility of expropriation, which the municipalities could have used in a number of cases (even though it was not necessarily used, the possibility that it might be employed gave power to the planning authority). Although *amtliche Umlegung* implies an infringement of ownership rights, the ownership in itself remains. Through the *Umlegung*, the land is replotted for the benefit of 'the society', but this is also in the interest of the landowners, because in return they get a plot with a higher value. Very often, because of this, the procedure evolves smoothly. However, the *Umlegung* is clearly not an act of self-determination by the landowner, and can in some cases work out like a compulsory purchase (for example for an owner of a small piece of land who receives only a financial compensation). If a landowner does not wish to participate, he can be made to do so by the government. But in principle, landowners receive a plot of land, which they can use as they wish. Their land is not purchased, but traded off for different plots of land.

One result of that is that the municipality – when it chooses to use an *Umlegung* – has to take account of the existing and future ownership structure, and cannot just draw the parcellation as it wishes. The owners have certain rights, one of which is to get a plot out of the procedure which is of the same size, and if not in the same, at least in a comparable position to the one they put in. As a result, negotiations between the city of Bonn and the landowners took place over the way the replotting was going to be carried out.

Ideas

The category of ideas as a vehicle to influence behaviour covers a large area of personal characteristics of the actors, mobilising concepts that allow coherence to be created between different actors, persuasive stories that allow actors to rouse the enthusiasm of others, etc. It is the least clearly defined category of power, and also the hardest to bring to light and analyse. Nevertheless, it can play a crucial role in the relations between the actors. Some of the case studies illustrate the diversity of forms in which ideas can shape the power and dependence relations in the development process, and hence influence its outcomes.

In the case of Bishop's Cleeve, the planning authorities – mainly the Borough Council in its role as controller of the development – used ideas about residential quality when trying to influence the house builders, by persuading them that they were responsible for some aspects of the residential environment.

In the case of Bois-Guillaume Portes de la Forêt, ideas played a bigger role in shaping the residential environment, through their influence in the development process. The area of the Portes de la Forêt is situated on a plateau north of Rouen. Normally, the rain water from this site would be drained into the river Seine. This would imply a large and costly installation of drain pipes leading the water down from the plateau towards the river. The private land developer proposed a solution in which the rain water was dealt with 'on-site'. It argued that this solution had another advantage – apart from the lower costs – in that it allowed the residential quality to be improved by offering an attractive green space.

This solution did not comply with the usual way of dealing with rain water in similar cases. But since decisions about urban development belonged to the competences of the mayor of Bois-Guillaume, he could choose another option than the standard one if he was convinced it was more suitable. That is what happened here. The technical and financial calculations with regard to the proposed system of rain water treatment

185

provided by the developer were judged sound enough to be adopted. Moreover, the possibility of arranging a green space combined with water in the plan area attracted the municipal council. So it decided to enable the land developer to realise this plan, which turned out to be satisfying for all parties involved. The innovative idea of the developer had become a binding element in the development process.

For the development process of Stuttgart Hausen-Fasanengarten to succeed in the form that it did, ideas were also of crucial importance. The GSL had a special place in the development process. As an independent party, it did not have any authoritative or allocative power. However, its role was crucial for the success of the operation. This was due to its communicative power. Although the municipality was also present in part of the negotiations, it was the task of the GSL to transmit the municipality's ideas to the landowners, and to convince them that participating was advantageous. The way this worked out is a good illustration of the binding force of ideas in the development process. The GSL did not have any resources to sustain its power in the interactions. The other parties did not really depend on the GSL for anything. Nevertheless, it did manage to convince all the parties to participate. It had only two means of power to achieve this. First, there was the power of conviction, a personal asset of the middleman. Second, it had the idea– coming from the municipality – of the *freiwillige Umlegung* which appeared to be the right idea at the right time and place to make the process work.

If the GSL had not managed to convince all the landowners to participate in the process under the given conditions, the process would have been seriously endangered. There would have been a way out if all the other landowners had accepted that special conditions applied to the landowners who did not agree with the initial conditions. But this would have been a fragile basis for an agreement. Another option would have been to carry out an *amtliche Umlegung*, just for the plots of those landowners who did not agree with the initial conditions. They would then have had to cooperate, but under conditions fixed by law (see case-study Bonn Ippendorf), so the flexibility of the *freiwillige Umlegung* would have been lost. Both solutions would not really have been satisfactory, which underlines the importance of the negotiations in which the parties sought an acceptable set of conditions for the *Umlegung*.

5 Decisions about the residential environment

5.1 The analysis of decision making

In this chapter, the focus is on decision making about the residential environment. The financial and institutional aspects of the development process reported in chapters three and four form the input for this chapter. The reconstruction of decision making in the development process shows how this is shaped by financial and institutional considerations together. Before engaging in this reconstruction, it is worthwhile to consider in more detail the notion of 'decision'. To this aim, we investigate first the theory of decision making in policy processes. This theory states that decisions cannot be ascribed to a single actor, but are made in interaction between the actors. We argue that the core of this observation, which is made in complex decision making processes, applies in general, but must be adapted for the study of decision making in operational decision making processes. Then the ideas on decision making are applied to the case studies.

The untraceable decision

The heading of this sub section comes from Muller and Seurel (1998: 101), who use it (in French: *introuvable décision*) in their study of public policy to introduce a section in which they deal with the problem of analysing decisions in policy processes. They argue that in policy processes, it is impossible to distinguish with precision the phases in which a system 'turns over'. The term 'turn over' refers to an event that makes the process enter into a new phase, a sudden change of the coordinates within which the decision making process proceeds. According to Muller and Seurel, it is not possible to describe such a turn over in terms of an individual 'decider', who can identify at a particular moment the different alternatives, and who takes a rational decision on the basis of an appreciation of these alternatives. This observation has important consequences for the possibilities and impossibilities of analysing decisions as an entity in policy processes.

An illuminating analysis of this subject is given by Teisman (1992 and 1997). He analyses the decision making about what he calls 'spatial investments', operations that change the spatial organisation (of a town, an area, a region) as a result of investment in physical structures (e.g. infrastructure, or real estate). The realisation of a housing scheme is a spatial investment. Like Muller and Seurel (1998), Teisman (1997: 23-26) argues that it is not possible to identify a particular moment in a policy process when a decision is taken. Instead, he proposes an alternative view on decisions – summarised in three central axioms – which might at first sight seem contrived, but turns out to be very enlightening.

In his first axiom, Teisman proposes considering decisions to be without a central point. As a second axiom, he proposes that decisions have no boundaries. In his view, it is not possible to distinguish between what belongs to a decision and what does not. Although this is an interesting perspective, the problems arise when it is used to analyse decision making processes. Teisman is aware of this, and that is why he completes his perspective on decisions by a third axiom, in which he states that each actor in the process demarcates what he considers to be part of the decision making process. Thus, each actor in his own way tries to make sense of the decision making process. This last axiom makes the first two axioms much less extreme. Instead of there being no central point in the decision making process, we can now say that there are as many central points as the actors in the process choose. This view on decision making is confirmed by empirical work. In various studies of policy processes, it appears that each actor makes his own reconstruction of the decision making process. The most renowned study in this respect is Allison's analysis of the Cuban missile crisis (1972). Studies made for example by Teisman (1992), Hajer (1995), or Zwanikken (2001) point from another perspective at the same phenomenon.

For the analysis of decisions, this has important consequences. It means that a decision is not an isolated act, but must be seen as a flow of activities and interactions in which there is no single moment or actor to which the decision can be accredited. From this observation, Muller and Seurel (1998: 103) draw the conclusion that with such a view on decisions, it is not opportune in the analysis of policy processes to look for a founding decision, nor to focus on who made the decision and why. The question is how the combination of perspectives and logics that are at work in a certain chain of events have led to the result that can be observed afterwards.

Complex decision making processes about spatial investment

The central assumption in Teisman's (1992) 'pluricentric perspective on decision making about spatial investments' is that decision making takes place in an environment of interdependence, in which no single actor is in charge (the concept of the interactive policy network, see section 1.2). To observe the decision making process in such an environment, Teisman (1992: 94-96) proposes a 'method of reconstruction of decision making'. This demands a focus on the interactions in the process. Teisman argues that three different types of interactions can be distinguished: initiation, adaptation, and selection. Initiation occurs throughout the development process. Each decision that provokes interactions about an item that was not previously the subject of interactions is characterised as initiation. Translated to this study, decisions about the design of the houses, the parcellation of the housing scheme, the level of servicing can all be taken as initiatives. This also means that all of the actors that take part in the process can initiate interactions by means of their decisions.

The interactions provoked by an initiating decision have as subject the adaption of that decision to make it acceptable for the other actors. Decisions are always part of a process of interaction, in which they are constantly adapted. In the analysis of decision making processes such as the housing development process, decisions can be discerned that are aimed at the adaptation of the initial decisions. This is the second category of decisions distinguished by Teisman.

On some occasions, out of this ongoing stream of decisions, one appears to influence the development in a direct way. This is the result of the process of selection that takes place in the interactions. The subsequent adaptations of the initial decision lead to a decision that is acceptable to all the actors involved, which is then selected. The term 'acceptable' that is used above needs some explanation. It is very possible that some of the actors involved do not, or do only partly agree with the selected decisions. That such decisions can be selected nevertheless, is a result of the power certain actors can have over others. When there is an asymetrical power balance, then actors can impose decisions upon others. But only to a certain extent: the parties are always interdependent, as a result of which no single actor can take a decision that infringes upon the activities of other actors without any support of the latter.

The observation of initiation, adaptation, and selection in decision making processes leads Teisman to the distinction of 'rounds' in this process. These rounds must not be seen as distinctive phases in the decision making process that occur in a chronological order. They are no more than an analytical tool for the description of the process. As such, the beginning and the end of the rounds is determined by the researcher. After a decision that the researcher judges to be crucial, a new round starts in which initiation, adaptation, and selection take place. The round lasts until a new important decision can be discerned.

Between 'complex' and 'operational' decision making processes

In order to use Teisman's view on decision making in this study, it needs to be adapted in some respects. The main reason is that Teisman is concerned with what he calls 'complex decision making processes'. Although these are not fundamentally different from what we will call 'operational' decision making processes, there is a gradual difference which is not negligible. The difference can be explained with an analogy to the difference between strategic plans and project plans as described by Mastop and Faludi (1997).

According to Mastop and Faludi the characteristics of strategic plans are that they '...deal with the coordination of a multitude of actors. Such coordination is a continuous concern. As all actors want to keep their options open, timing is of central importance. Rather than a finished product, a strategic plan is a fleeting record of agreements reached. It forms a frame of reference and is indicative. The future remains open' (1997: 819). As opposed to strategic plans, project plans are described by

Mastop and Faludi as plans that '...provide blueprints of the intended end-state of the physical environment, including measures necessary to reach that state. The only important social interaction is when the plan is being adopted. Thereafter the plan forms an unambiguous guide to action precisely because the measures to be taken are routine so that we "know" the outcomes' (Ibid: 819).

In spite of the differences between both types of plans, the basic aim is always the same, namely to guide future decisions and measures. According to the kind of problems at which the future decisions and measures are aimed, the plans vary. On the one side, there are complex problems. These are unique, they involve a large number of actors, and they are open so that all the time new actors can get involved or actors already involved can disappear. Examples of this kind of problems are the choice for the route of a new railway (Teisman, 1997), or the drawing up of a circulation plan for an entire city (Flyvbjerg, 1998). On the other side, there are more 'routine' problems which are clearly defined, for which the actions are routine, and the outcomes certain. Examples of this kind of problem are the construction of a house, or a boat.

And now we come to our analogy with decision making. As plans are made to guide decision making, they concern the same problems as the decision making itself. The distinction between 'routine' and 'unique' problems used by Mastop and Faludi to distinguish between types of plans can therefore be used to distinguish between types of decision making also. Housing development processes as we describe them, have characteristics of both 'unique' and 'routine' problems, and hence of complex and operational decisicion making.

Describing the housing development process as a 'routine' problem is not appropriate. An important element from Teisman and from Muller and Seurel that we want to incorporate in our analysis is the appreciation that decisions are not taken by one actor, but emerge in an ongoing stream of interactions. That also implies that decisions are not taken once and for all, but are used in the continuation of the process. This cannot be taken into account when the housing development process is described as a 'routine' problem that could be solved using a project plan.

Describing the housing development process solely as a 'unique' problem is not appropriate either. During the process of housing development, a number of 'points of no return' occur. With this, we refer to moments when a decision is taken because it is the right time for it in the process. There is a difference and a similarity between these points and Teisman's rounds of decision making, in which initiation, adaptation, and selection occur. The difference is that our points of no return are linked to specific moments in the process. Housing development processes are 'routine' enough to distinguish a certain pattern of decisions that have direct consequences 'on the ground' and hence for the process. The similarity is that this decision can be changed or reviewed in subsequent phases of the process. However, this usually requires an important effort, because of the results 'on the ground' that the decision has provoked. In the case descriptions in section 5.2, the idea of rounds of decision making is combined with the idea of points of no return through the activities in the development process that we have discerned in chapters three and four. We recall them here:

−   identification of development opportiunities;
−   land assembly;
−   land development;
−   housing production;
−   final ownership/use.

In chapter three, in the financial analysis, we have described how a financial margin in the development process can occur during each of these activities. In chapter four, in the institutional analysis, we have used the activities to describe the actors, the roles, and the power relations in the housing development process. By relating the decisions that were taken in the case studies to these activities, we can now link the financial and the institutional analysis. In section 5.2, we describe the interactions in the cases with the use of the perspective on decision making developed above. In section 5.3, we zoom in on how the decisions made in the housing development process influenced selected aspects of the residential environment. In section 5.4 the findings of this chapter are summarised.

5.2   Case study files: interactions and decisions

Arnhem Rijkerswoerd

The interactions in Rijkerswoerd were dominated by the municipality. It decided when and where development was to take place, and who was to do what. The discussion about the expenditure on the residential environment was largely an internal process within the municipality. It took partly place during the making of the *bestemmingsplan* (local land use plan). In this phase, finance started to play a role. The people who worked on the plan knew that it had to be realised, and that the costs of realisation should not exceed the income from the sale of the land. Moreover, it is a legal requirement that each *bestemmingsplan* is accompanied by an *exploitatieopzet*, a financial account in which the municipality shows how it balances costs and income. So in the first instance, the interactions over the use of a possible financial margin took place within the confines of the municipal organisation.

The income of the municipality was not high enough to cover the costs of realisation of the plan. For that reason, it applied to central government for a *locatiesubsidie*. This subsidy came with certain conditions regarding the residential environment. So at this stage of the development process, a first round of interactions over the residential environment started, between central government and the municipality. The outcome was a set of requirements, to be used as standards for the realisation of the housing scheme.

In the next phase, the *bestemmingsplan* was made. The requirements of central government were incorporated into the plan. This plan was delibertately not very detailed. The municipality needed other parties – housing corporations and private house builders – to realise the housing scheme. To get these parties interested in building on the site, their wishes regarding where to build which type of housing had to be taken into account. The private developers as well as the housing corporations used this opportunity to try and realise a housing scheme that corresponded as much as possible with their objectives.

On some occasions, private developers did not reach an agreement with the municipality about the way in which a particular sector was to be developed. The private developers were sometimes not able, or did not want to comply with the conditions of the municipality. In those cases, the municipality sometimes chose to use its power as landowner and did not sell the building land to the developer. However, this was a painful thing to do, because a lot of work had already been invested in a plan before the parties came to such a decision. So it meant a significant loss of time and money for each of the parties. This did not happen frequently because the developers were generally quite satisfied to be able to build. The market was good during the development of Rijkerswoerd phase two, as a result of which the developers had more 'room to play with' in their assessment of the costs and benefits of different house types, design, and servicing levels.

This brought the municipality in a very comfortable position. It could define the margin for negotiation over the exact form the development (although this had to be within the requirements of the *locatiesubsidie*). During the interactions, the municipality was the director: If it did not reach an agreement with the developer, it could decide not to dispose of the land, with the result that the developer remained empty handed. The only room to play that remained for the developers was the exact design of the houses they realised, and to some extent the street pattern and parcellation, because only the main roads were fixed by the municipality.

Figure 5.1   The main through road with footpath and cycleway in Arnhem Rijkerswoerd

Table 5.1   Decisions in Arnhem Rijkerswoerd

Identification of development opportunities	- Choice of size of the area of land to be acquired - Choice for active policy of land acquisition, and hence of directing role for the municipality
Land assembly	- Application for *Locatiesubsidie* to central government combined with requirements regarding number and tenure of dwellings
Land development	- Selection of developers: private house builders and housing corporations - Space for facilities designated in the local plan - Parcellation, road layout, number of houses, and public spaces fixed in the local plan - Design guidelines fixed in the local plan
Housing development	- Architectural design of the houses - Exact numbers of market/social sector dwellings to be realised
Final ownership/use	- Commercial and public facilities choose to settle in Rijkerswoerd

A large part of the interactions between the parties involved in the development process of Oldenelerbroek took place within the project teams (see section 2.4). However, before the parties met in those teams, the municipality had already fixed a large number of preconditions, for example regarding road layout, parcellation, housing type, but also concerning design aspects. These were presented to the other actors in a plan document. Then the housing developer – either private house builders or housing corporations – of each section of the housing scheme presented his plan, taking into account the guidelines and conditions given by the municipality. In principle, the only thing they needed to do was to design the houses, since the rest of the plan was already designed by the municipality. However, the plan of the municipality did not always please the developers, in which case they had some margin to manoeuvre, as long as the changes to the plan they proposed did not change the basic principles laid down by the municipality. The land was transferred only when an agreement was reached about exactly how to develop the section.

Although it admitted a certain margin to manoeuvre to the developers, the municipality's influence was considerable. First, the municipality provided a plan with the conditions that were an input in the interactions with the developers. The developers then reacted and the result of the interactions was usually a slightly adapted plan for the design of the area. But before this 'new' plan was approved by the municipality and a building permission could be granted, there was another examination, by the independent *welstandscommissie* (committee for design).

The way in which this worked out in practice can be illustrated with the case of an appartment building of six floors that the municipal plan projected as a landmark on the edge of a central open space. Because of the central place of this building, the municipal urban designer had some requirements as to the colour and the form of the building. However, the architect of the housing corporation that was to develop it did not agree with the choice of colour. Neither did he agree with the proposed form of the building. So the municipality's urban designer and the housing corporation's architect started to interact, through the intermediary of their project leaders. From these interactions resulted a building that had another

196

form and another colour than what the municipality initially proposed, but upon which both parties could agree. This did not mean that the building could start, because first the independent committee for design had to judge whether or not this newly designed building fitted into the overall picture of the development. Only after this committee had given its approval, could the construction work start.

Apart from the design, interactions also took place about the prices that were calculated. Land for social housing was sold at a lower price than land for market sector housing. This sometimes led to some controversy when a housing corporation also built houses in the market sector. At a certain stage in the development process, the municipality made the decision that more market sector housing could be built in the plan area. For a housing corporation this meant that instead of building only social sector housing, it could also build some market sector houses.

Figure 5.2   Centrally located apartment building for social
             housing in the design agreed between housing
             corporation and municipality (Zwolle Oldenelerbroek)

197

The housing corporation was used to receiving the land for a certain –
reduced – price. It started its calculations on the basis of this price. This
allowed it to make some profits on the market sector houses, which could
be used to improve the quality of the social housing in the rest of the
section. To the municipality, however, this would mean implicitly –
through a reduced land price – subsidising market sector housing. For the
market sector plots, the municipality demanded the same price from the
housing corporation as from private house builders. From the
municipality's point of view this was only fair. In the eyes of the
corporation, the municipality now received a profit on the land that the
corporation could have used very well for an improvement of the quality of
the social housing.

Although, most of the time, the municipality is described as a single
actor, it consists of several departments, which sometimes had different
interests and objectives. Differences in objectives became apparent for
example when the department of financial affairs decided to ask a higher
price for the land on which the housing corporation was going to build
market sector houses. From the point of view of this department, this was a
logical decision. The people at the department of housing, however,
reacted more like the housing corporations. They realised that this decision
might have consequences for the level of the rent the housing corporations
would ask. However, the municipality succeeded in managing these kinds
of controversies internally. Towards the other actors, the municipality –
represented by the project leader – spoke with 'one voice'.

Table 5.2   Decisions in Zwolle Oldenelerbroek

Identification of development opportunities	- Size of the area, general road structure, and appreciation of number of houses to be developed presented in global local land use plan - Choice for an active land policy of land acquisition and hence for a directing role of the municipality
Land assembly	- Number of houses, road structure, parcellation, public spaces fixed in the elaboration of the local land use plan - Design guidelines fixed in local plan
Land development	- Selection of developers: private house builders and housing corporations - Parcellation, road structure, and public spaces slightly adapted in negotiations with developers
Housing development	- Architectural design of the houses, within design guidelines issued by the municipality - Exact number of market and social sector houses determined
Final ownership/use	- No important decisions concerning the residential environment

Bishop's Cleeve

At the beginning of the process, interaction – in the form of negotiations – took place in the preparation of the private agreements between planning authorities and house builders. In these negotiations a lot of things were fixed that were of direct importance for the residential environment, for example with regard to open space. There was not much open space in Bishop's Cleeve, so the Parish Council and the Borough Council wanted to use the new housing development to create more. They asked the house builders to provide open space within the housing development of Bishop's Cleeve. The house builders preferred to offer more land that was not in the new development but situated in the green belt around Cheltenham. For that reason the chance that this land would ever be developed as housing land was very small, so it was of less value to them. Because of this, the planning authorithies even managed to obtain money from the house builders for developing this area as a green space.

199

With regard to the bypass, the house builders also changed the plans of the planning authority, but this time they used a more formalised negotiation path. It is an example that had a great impact on the whole appearance of the area. What happened? The District Council proposed a bypass in the first local plan. The house builders did not agree with it and went to appeal. On the basis of their arguments, they won the case and the bypass was realised in the way they had proposed it. It meant that the road they had to provide was much shorter (and therefore cheaper) than it would have been in the first place. Moreover, it meant that they were able to develop more of the land they had acquired for housing. So financially it was very interesting for the developers. This was not an argument that would count as a material consideration in an appeal, nor did the developers use it as an argument. They won the case on the basis of other material considerations that were judged valuable by the inspector.

Figure 5.3   A "cul-de-sac" in Bishop's Cleeve

This overruled some considerations of the District Council, the most important of which was that the housing development should not 'cross' the bypass. The bypass would thus form a 'hard' border between the urbanised area and the green fields on the other side. In the new plan, the

200

house builders were allowed to build on both sides of the bypass. This made the border between the village and the countryside more vague, which the District Council felt would make it more difficult in the future to stop further housing development.

The cases described above were 'big events' in the interactions. The 'day-to-day' development control also involved a considerable amount of negotiation, for example about the amount of open space the house builders had to provide, or about the design of the housing schemes. The planning authorities had their statutory powers to influence the behaviour of the house builders. However, in these negotiations, they had to do a lot by persuasion. An example of this is the way in which they tried to realise a footpath and cycleway network in Bishop's Cleeve. Planning officers had to speak with the house builders and persuade them that this was a good thing to do. Hence, a lot depended on the persons involved.

Whenever a house builder applied to the Borough Council for a building permission for a specific project, this application was checked against the standards of the Borough Council. This implied new interactions. In fact, sometimes the preparation of a planning application already involved negotiations between the house builders and the planning authorities. The house builders put in their outline application and then the planning authorities reacted to that. They said what they thought of the proposed development and both parties tried to find a solution that was suitable for all parties. Usually, the planning authorities referred to this activity as persuading the house builders. But everybody probably had their strategy in the negotiations, which made it difficult to discern afterwards who persuaded whom.

Table 5.3   Decisions in Bishop's Cleeve

Identification of development opportunities	- House builders decide to buy land - Planning authorities designate plan area (size) and number of houses to be developed
Land assembly	- Plan area (location) determined by the inspector in reaction to an appeal by house builders
Land development	- Transfer of land for school and park (plus money for development) by house builders to planning authority - Parcellation and road layout
Housing development	- Exact number of houses developed (considerably higher than number designated in local plan), and hence higher housing density - Architectural design of the houses
Final ownership/use	- Handing over of public spaces to planning authority after period of maintenance by house builders

Cramlington North-East Sector

An important round of interactions in the development of Cramlington concerned the 1974 agreement. In this document, both parties agreed what their contribution to the development would be. The reason for the planning authority to sign the agreement was obvious: it was in its direct interest to try and reach an agreement with the house builders because it allowed it to realise its objectives of a good quality housing scheme. For the house builders, signing the agreement was more a negative choice. They wanted to build houses, and to make sure that they were able to do so, they were prepared to sign an agreement that obliged them to spend some money on things in which they were not primarily interested. And of course, they knew that in a development of this size there would have to be a basic level of services to interest people to buy the houses they were building.

The district council did not issue quality standards in the form of design guidance for the housing development in Cramlington. It did have some influence on the design because when a planning application came in, it could – by persuasion – try to influence the design in the way it wanted. Apart from this influence by persuasion, the district council applied the

202

minimum requirements regarding certain aspects of the residential environment that were standard throughout the country. The house builders had to comply with these standards. But they assured only a minimum level of residential quality. Expenditure on the residential environment above this minimum level was subject to negotiations. Sometimes this worked, but the house builders could choose not to comply with what was proposed by the District Council, or to apply it in in their own way. This happened with the 'special design sites'. The house builders did not like to build on these sites because it was unusual and they were not sure that the people would like what was offered. They preferred to build houses of which they knew that they would be able to sell them easily.

The realisation of facilities posed problems because everything was arranged in negotiations between the two house builders and the local planning authority. Some commercial facilities like shops had to be realised by third parties, not involved in the agreement. But since the land was owned by the house builders, this did not always work smoothly, which sometimes caused problems for the overall development of the scheme. The house builders received an outline consent for shops on a particular site, and then they had to market them – together with the planning authority, because it was equally in its interest that the shops were realised. The house builders sold the land to a retail developer only if they received a commercial price for it. Therefore it took time before the facilities came: They had to be provided by private parties who would only build when it was economically viable, that is, when there were enough people living in the area. But people wanted to buy houses only if there were facilities nearby. And house builders wanted to build houses only when they knew that they were able to sell them. This created a sort of vicious circle which in North-East Cramlington caused some 'hiccups' in the process, as the respondents put it.

Figure 5.4   Standard housetypes with different finishings in Cramlington North-East Sector

Often, changes in the context led to new negotiations. This happened for example as regards the provision of open space and the planning of schools. Because of the changing expectations about the expected number of inhabitants, fewer school sites were necessary than had been planned. There were negotiations about what to do with these sites, the result of which was that they became partly green spaces, and partly a housing area.

Table 5.4   Decisions in Cramlington North-East Sector

Identification of development opportunities	- Land acquisition by house builders - Choice to develop the area as a new town by private house builders - Size of area and number of houses fixed in C.D.A plan.
Land assembly	- (Probably) lobbying of planning authorities by house builders for the development of the area
Land development	- Parcellation, road layout, and public spaces realised in final form - Local planning authority issues guidelines for housing density and special design sites
Housing development	- Number and type of houses actually realised - Architectural design of the houses
Final ownership/use	- Handing over of public spaces to planning authority after maintenance period by the house builders - Commercial facilities choose to settle in Cramlington

Bonn Ippendorf

An important part of the interactions between the actors in the development process of Ippendorf took place through official documents, such as the land use plan and the plan for the *Umlegung*. But direct interactions through negotiations between the municipality and the landowners also occurred. However, the type of development process chosen here did not leave much space for this. Most aspects of the process were arranged in official plans and procedures.

The *Bürgerbeteiligung* (public inquiry) was a phase in the preparation of a local land use plan in which the citizens could give their opinion on the plan proposals of the municipality. The remarks made at this stage were input in the following round of planning within the administration. In some phases of the local planning procedure, the interactions between the municipal administration and other parties concerned with the plan are obvious, for example in the preparatory phase, between different levels of government, or in the phases of public participation. Describing the *Bebauungsplan* in itself as a vehicle for interaction may seem less obvious. Yet the plan can be seen in this way. In the land use plan the municipality prescribed certain design guidelines for the houses. The land use plan was the medium through which these prescriptions were communicated. The house builders responded when applying for building permission, and in that sense there were interactions. But these were formalised, indirect interactions (because they took place through the intermediary of the planning procedure).

Figure 5.5   Houses in varying stages of construction in Bonn Ippendorf

However, these were not the only interactions in the development process. To carry out the *amtliche Umlegung*, direct interactions between the municipal administration and the landowners were required, in order to realise a realistic replotting of the area. From the point of view of the municipality, these interactions were part of the procedure of the *Umlegung*. The municipality had to draw up a new parcellation, and for this it needed to reach an agreement with the landowners. Within the municipal administration, the procedure of the *Umlegung*, and that of the *Bebauungsplan* needed to be adjusted to each other. The department of land registry had to draw up a new parcellation in which each of the landowners would receive his share, and that corresponded to the development proposals specified in the land use plan. This required clear internal communication within the municipality, which was ensured by drawing up the plan and the parcellation in close cooperation within the same municipal department.

Table 5.5   Decisions in Bonn Ippendorf

Identification of development opportunities	- Choice for the procedure of the *amtliche Baulandumlegung* - Determination of size of plan area
Land assembly	- Parcellation, public spaces and road layout fixed in *Umlegungsplan* and *Bebauungsplan* - Number and size of building plots and design guidelines for houses fixed in *Bebauungsplan*
Land development	- Land for roads and public spaces claimed by municipality from first landowners (as part of procedure of *Umlegung*)
Housing development	- Architectural design of the houses (taking into account design guidelines) - Number of houses on the site, and hence housing density
Final ownership/use	- Arrangement of public spaces

Stuttgart Hausen-Fasanengarten

Interactions played a crucial role in the initial phases of the development process of Hausen-Fasanengarten. Whether or not the operation would be realised depended on good communication about its potential advantages, and the conditions that were to be taken into account. To ensure the success of this communication, an independent agent, the GSL, was appointed. Its first task was to convince the landowners to enter into the *freiwillige Umlegung*. The only argument for that was the value increase of the land due to the land conversion. Because the operation could only proceed if all the landowners agreed, the GSL paid much attention to the way in which it put forward this argument. It was very important that all the landowners were treated on the same basis. If one of them had found out that another received more for his land, this might have endangered the whole procedure.

In the first round of negotiations, in which the municipality and the landowners had to agree on the conditions for the development (the *Umlegungsmaßstab*), the municipality adapted its first requirements. In the beginning it wanted to receive 36% of the land to use for roads, public spaces, etc. During the process, it decided that 30% of the land for public spaces was sufficient. The main obligation the municipality took on was to realise the plan within a time span of seven years. If not, the landowners could claim back the money from the municipality for the part of the plan that had not been realised. In return, the landowners were obliged once the building plots were ready, to build on them within five years.

But the most noticeable obligation of the landowners was that they took care of the land development and the servicing of the area. To this aim, they formed an *Erschließungsgemeinschaft* (community for land development) which was directed and represented by the GSL. On behalf of the *Erschließungsgemeinschaft*, the GSL signed an agreement with the municipality in which all the requirements regarding the level of servicing of the area were fixed. Thus, the municipality ensured that the work would be done according to its quality standards.

It was the task of the GSL to convince the landowners of the advantages of carrying out the land development. It used three arguments. First, the transparency of the whole operation: the landowners could follow the whole calculation that was made about the costs of the land development, and they knew that if they would pay this now, they would not have any claims later on. The second argument was that the GSL as a representative of the landowners could carry out the land development cheaper than the municipality. Its overhead costs were lower, and as a private party, requirements of procedures to be taken into account were less strict. Finally, the GSL could organise the whole land development faster than the municipality, for much the same reasons. This meant that the time interval between the start of the process and the moment the landowners could capture the value increase of their land was reduced.

For the intermediary role of the GSL, the interactions with the municipality were as important as those with the landowners. Stuttgart is a

208

big city with a lot of competence in its organisation, so a lot of coordination within the municipal organisation was required to succesfully complete such a complex operation. The municipality wanted to have an influence on the form of the plan. To this aim, it used the official local planning procedure. For example, it specified requirements regarding the design of the area. The municipality had made an outline design for the whole plan. This outline was presented with the local plan in the municipal council and was agreed upon. As a result, the plan had some status to direct the actual construction of the area, although it did not fix the end state of the development.

The municipality also used the local plan to influence the type of housing to be built in the area. Because the municipality wanted the houses in the area to be accessible not just for higher income groups, it fixed a maximum floorspace for some of the houses. The plan reserved space for shops and services also. Here too, the parcellation in the *Umlegungsplan* was adjusted to the local plan: The reserved spaces for shops and services corresponded to parcels of land owned by the municipality. Thus, the municipality had some influence on the realisation of these facilities.

Table 5.6   Decisions in Stuttgart Hausen-Fasanengarten

Identification of development opportunities	- Choice for the procedure of an *Umlegung mit freiwillig vereinbarten Konditionen* - Choice to make the landowners carry out the development work - Size of area to be developed with social housing (20% of land sold at reduced price as *Sozialbeitrag* to municipality)
Land assembly	- Parcellation, public spaces, and road layout fixed in *Umlegungsplan* and *Bebauungsplan*
Land development	- Supervision on standards of servicing by municipality
Housing development	- Transfer of land for facilities to municipality - Sale of land for reduced price (*Sozialbeitrag*) to municipality - Exact number of market sector and social sector houses to be developed - Architectural design of the houses
Final ownership/use	- Commercial facilities choose to settle in Hausen-Fasanengarten

The procedure of the *Zone d'Aménagement Concerté (ZAC)* determined the space for the interactions in Portes de la Forêt. By indicating the area of Portes de la Forêt as a ZAC, the municipality created a negotiation situation that constituted the framework for further interactions in the development process. Once the developer for the area was chosen, in a competition that was held between two competing developers, the procedure of the ZAC prescribed a sequence of steps that had to be followed in the process. After the municipality had delimited the area, the developer proposed a plan for the development. His proposals were discussed with the municipality. The latter checked whether the plan corresponded with its objectives and its policy. This process of designing and checking the plans for the area took place in close cooperation between the developer and the municipality. But because these negotiations took place 'behind closed doors' it is not possible to give an elaborate description of the exchanges that took place. It is possible, however, to reconstruct the tenor of the talks.

The developer knew that the municipality was prepared to sell the land at the rather moderate price for which it had been acquired. The municipality replaced the higher price that it could have received for the land by contributions from the developer towards the residential environment, and towards public facilities. That was the basis for the plan proposed by the developer. In the plan, some elements were introduced that combined a positive value for the residential environment with relatively low costs (the system for on-site storage of rain water and the hierarchical system of roads, with mixed traffic streets to serve small clusters of houses). The framework of the ZAC, in which all aspects of the development were negotiable, made this possible. If the municipality had just judged the proposals of the developer on standard criteria, it would have been difficult to realise these innovative aspects because they did not match with current standards.

The negotiations led to a plan that was officially laid down in the *Plan d'Aménagement de Zone* (PAZ). This was the official planning document that was produced within the procedure of a ZAC, and replaced the local land use plan (*Plan d'Occupation du Sol* or POS) for the area concerned.

The proposal for a PAZ then went into a public inquiry. The publication of the PAZ marked the end of the negotiations, and fixed the end-results. From this moment on, further changes in the plans would require modification of the PAZ, which would involve procedural requirements such as a new public inquiry. In addition to the PAZ, the results of the negotiations were laid down in a convention signed by the municipality and the developer. In this convention, all the financial conditions were laid down, and the obligations of the two parties towards each otherweare appointed. This convention was signed only once the PAZ was officially approved.

The municipality hardly interacted directly with the buyers of the plots. This role was conceded to the private developer. Some aspects of these interactions were of consequence for the end result of the process. The developer imposed architectural requirements upon the house builders. To be able to do so, the developer claimed to have moderated the prices of the building plots to compensate for the costs of the architectural requirements to the house builders, and for the constraint upon their freedom to build what they wanted. The municipality did have direct interactions with the first landowners. These were crucial for the development process. Because the municipality managed to acquire the land for a low price, a margin was created for investment in the residential environment. The importance of the low price of the land for the realisation of the development was stressed by both the municipality and the developer.

Figure 5.6   On-site storage of rain water arranged as an attractive green space in Bois-Guillaume Portes de a Forêt

Table 5.7   Decisions in Bois-Guillaume Portes de la Forêt

Identification of development opportunities	- Choice by municipality to pursue an active policy of land acquisition - *Grandes options directrices* (general outlines of development) are fixed (main road structure, number of dwellings, size of site) - Choice to use the procedure of a ZAC
Land assembly	- Choice to have the actual development carried out by a private land developer - The land developer is chosen by means of a contest
Land development	- Parcellation, public spaces, road layout and number and type of houses fixed in local plan (PAZ) - Design guidelines are fixed in the local plan - Contribution by developer (financial and in kind) towards public facilities and green spaces is agreed - Land for school is transferred to the municipality
Housing development	- Architectural design of houses within design guidelines issued by municipality
Final ownership/use	- Public spaces are handed over to the municipality, after maintenance period by land developer - Commercial facilities choose to settle in the area

Rennes la Poterie

During the development process there were constant close interactions between the SEMAEB and the municipality. These took place within the framework of the ZAC-procedure. In the first phase, the phase of creation of the ZAC, both parties together defined the area of the plan, a programme of what was going to be developed, and a financial account. On this basis, the second phase, that of realisation of the development, was entered. The SEMAEB being the developer of the area, took care of the preliminary studies and officially drew up the plan. Because the municipality of Rennes possessed the necessary knowledge within its organisation, it took an active part in the planning, which was done in *concertation* (consultation), reflected in the name *Zone d'Aménagement Concerté*. The special role of the SEMAEB was to check whether the policy aims of the municipality – the programmes of housing and of public facilities, the level of servicing of the area, etc. – were financially feasible, and if not to propose alternatives. For example, if the municipality had wanted to realise more social housing, this would have meant that the building land would be sold at a lower price. In turn, that would have meant lower incomes for the SEMAEB. With the agreed financial accounts in the hand, the latter would then have required a higher contribution from the municipality to the development.

During the phase of creation, the interactions between the SEMAEB and the municipality mainly took the form of negotiations. In the next phase, that of realisation, negotiations still played a role. However, they became less important. An important part of the interactions consisted of a more routine supervision of the SEMAEB by the municipality. To this aim, the SEMAEB was officially obliged to discuss the proceedings of the operation once a year with the municipality. But in practice, whenever new developments occured – either about finance or concerning design or planning – this was discussed with the municipality. All deviations from the plan that had initially been agreed were thus validated by the muncipality, or else abandoned. When changes in the context occurred, sometimes renegotiation of earlier fixed conditions was required. However, the room for such renegotiations was limited: The limits were fixed in the *Plan d'Aménagement de Zone* which was the result of the negotiations in the phase of creation. Changes to this plan required an official, legal procedure.

In the course of the development process, there was also a continuous interaction between the two central actors, the SEMAEB and the municipality, and the house building companies. It was in the house builders' interest to limit the supply of dwellings. This allowed them to sell their houses quickly, and for good prices. However, to be able to build houses, they depended on the SEMAEB and the municipality for building land, the necessary connection to networks, etc. Moreover, the municipality had the responsibility for the granting of building permissions. Its objective was that there should always be enough dwellings brought onto the market so that the prices would not rise too much. It wanted to avoid a shortage of housing. But the municipality did not want to build houses itself. The SEMAEB had balanced the budget of the operation on the basis of an estimated area of building land to be sold each year, at an estimated price. The number of dwellings built had repercussions on these estimates, hence on the balance sheet of the operation. For these reasons, all the parties had to come to an agreement as to how many dwellings should be brought onto the market each year. This led to constant interactions between them during the development process.

Table 5.8   Decisions in Rennes la Poterie

Identification of development opportunities	- Choice for an active policy of land acquisition by the municipality - Choice to leave the actual development to the SEMAEB - Designation of area to be developed and number of houses
Land assembly	- No important decisions concerning the residential environment (took place before any concrete plans were made)
Land development	- Parcellation, road layout and public spaces are fixed in local plan - Space for facilities os designated and developed as such
Housing development	- Architectural design of the houses - Exact number of houses in market sector and in social sector
Final ownership/use	- Handing over of the public spaces to the municipality (after maintenance period by SEMAEB) - Commercial facilities choose to settle in La Poterie

5.3   Realising the residential environment

As described in chapter two, comparing the level of residential quality in different countries is very difficult, if not impossible due to cultural differences. However, it is possible to distinguish items that are judged in all cases as influencing the residential environment. For this study it is also important that these items are influenced during the development process. The combination of these two criteria has led to the selection of four items for which the decision making process is reconstructed below. These items are: housing density, urban design, public facilities, and the mix of social/market sector dwellings. Although each case has its particularities, our focus is on the more structural, recurring aspects of decision making about each particular item.

Housing density

In most cases, no direct decisions are taken about housing density. The density of the housing scheme is usually a result of decisions about the parcellation and the design of the area. The main influence on the 'selected' parcellation is exercised by the actor who is responsible for the land conversion. The parcellation this actor initially draws up is a result of his objectives, and of the constraints imposed by norms and standards, by the characteristics of the site, and by the situation on the market for housing and housing land. Whether it is a public or a private actor who draws up the parcellation does not fundamentally change this. Often, an initial decision about housing density is made by a public actor who prescribes the number of houses to be realised in the plan area. Only in the case of Bonn Ippendorf, was there no public decision preceding the development about the number of houses to be realised on the site. What happens to this initial decision depends on the type of process. In the cases in the United Kingdom, this number of houses was used only to determine the size of the site to be developed in a general way. In the other cases, this number of houses formed a more or less 'hard' input by the municipality or even by central government.

In Arnhem Rijkerswoerd the municipality as a landowner drew up the parcellation of the housing scheme, initially based on a total number of 3000 dwellings. At the moment the municipality applied to the central government for *locatiesubsidie* this first plan became untenable. The municipality had to respect the fact that the central government now also played a role in the development process by imposing its standards. An important standard from central government was the total number of houses that could be realised in the housing scheme. If these standards were to be applied, 7000 houses had to be realised in the area of Rijkerswoerd. However, another central government policy played a role. The growth centre policy, launched in the *Derde Nota Ruimtelijke Ordening* (Third Report on Spatial Planning) prescribed that the necessary houses in the region were to be built not in Arnhem, but in the adjacent municipalities of Duiven and Westervoort. The municipality of Arnhem, having already acquired the land for Rijkerswoerd, objected to this consequence of the growth centre policy and a round of negotiations between the municipalities of Arnhem, Duiven, Westervoort and central government started, in which the plan to realise 7000 houses in the area was again adapted. The result was that Arnhem was allowed to build 4500 houses in Rijkerswoerd.

The negotiations that took place in relation to the *locatiesubsidie* and the growth centre policy had a substantial influence on the eventual housing density in Rijkerswoerd. The end result was more than the initial 3000 proposed by the municipality, because central government required a higher density to grant its subsidy. However, it was less than the 7000 houses that were the central government's first proposal when it applied its criteria for granting a *locatiesubsidie* to the area. Because of the growth centre policy, part of these houses were redirected to the neighbouring municipalities. So in this case, 'external factors' greatly influenced the number of houses to be realised on the site, hence the housing density.

The link between the decision making process about the number of houses and the financial accounts goes through the *locatiesubsidie*. When making its plan, the municipality knew that if it did not manage to balance costs and income, it could apply to central government to bridge the gap. But to do so central government prescribed the number of dwellings, hence the housing density to be realised. This was partly an economic consideration. The proposed density was one that was feasible without excessive costs caused by either extremely low (e.g. length of networks in relation to number of connections), or very high densities (e.g. technical problems to accommodate traffic). With the proposed density a reasonable income could be earned from the sale of the plots, which would reduce the amount of subsidy to be given. Central government was not going to subsidise the realisation of a 'luxurious' housing scheme. Its subsidy was aimed at enabling municipalities to provide sufficient housing to suit the need, not more.

Bishop's Cleeve is probably the example where the influence of either local or central government on the housing density was the most restricted. The development started with a plan by Gloucestershire County Council to develop 1000 houses in Bishop's Cleeve. This number was taken over by

Tewkesbury Borough Council when they made their more detailed plan for the housing scheme. However, the development itself was carried out by private house builders, and they had their own ideas about density, which turned out to be much higher than the numbers proposed by the local planning authorities (around 1700 instead of 1000 houses). How did this happen?

The proposal by the local planning authorities to realise 1000 dwellings in Bishop's Cleeve was the basis for the determination of the area to be designated for future housing development. That implied a certain idea about a desireable housing density. At the same time, the local planning authority knew that it was not going to realise the scheme on its own, but that private house builders would do that. And the behaviour of these private house builders was driven by economic considerations about costs and income and supply and demand. In other words, the proposed density had to be feasible from an economic point of view. So the initial proposal of 1000 houses made by the local planning authorities was not solely based on an image of the desired residential quality, but also on an idea about what density was realistic in terms of costs and income. The house builders seemed to agree with this. When they objected to the plan, it was not against the number of dwellings or against the density to which this would lead when realised in the designated area. They proposed an alternative layout of the plan and especially a different routing of the bypass.

In the course of the development process the number of houses realised turned out to be over two thirds higher than the original proposal. The reason for this was that the house builders adapted their housing supply to the market demand, and in the period over which the housing scheme was realised, apparently there was a demand for houses on smaller plots, or in any case for more houses. Building more houses on the same site did not necessarily mean that the income rose, because higher densities also meant more building costs, higher costs of land development, and higher supervision and marketing costs. A representative of the house builders in Bishop's Cleeve claimed that in general it is more profitable to build bigger houses. However, whether this is also the case if the total area to be built upon is limited – as it was in Bishop's Cleeve – can be questioned. The assumption that the main reason for the house builder to build more houses on the site was to optimise their income seems fair, although due to the confidentiality of this kind of considerations, it is no more than an assumption. Whatever the reason, the result was that both the borough council and the inhabitants of Bishop's Cleeve were confronted with a much bigger housing scheme than they had originally expected on the basis of the planning documents.

The borough council did not have much power to influence this density. The permission for residential development that was granted did not include any requirements as to maximum (or minimum) densities. Therefore, the borough council could not use any public powers. The only way to try and influence the house builders on this subject was by trying to persuade them. Although this did not work very well in terms of the density, it might have worked in another way. The argument that many more residents had to be accomodated in the plan area because of the higher density of houses, was used by the Borough Council officer to persuade the house builders to provide additional land for open spaces. As we will see below, they did provide this land.

In Cramlington North-East Sector, the planning authority tried actively to influence the residential density, but as in Bishop's Cleeve without much success. In 1982, a new document was issued by the 'Planning and development services committee' of Blyth Valley District Council, concerning the development of the North-East Sector of Cramlington. This document, called the 'Cramlington North East sector housing strategy' was an addition to the original Comprehensive Development Area (C.D.A.) plan. It was a result of discussions held by the committee with the developers. 'The purpose of the strategy is to provide general guidelines for the development of the remaining areas of housing land in the North-East sector.' Behind the strategy that is presented '... is the need to ensure that there is an adequate supply of housing land, giving continuity of building for the various development agencies, and that there is a reasonable choice of attractively designed housing' (Blyth Valley District Council, 1982). The prescription of housing densities in this document did not have much effect in practice. The District Council had no means to impose housing densities on the developers. It could only indicate land for housing (what it did in the C.D.A. plan). It was up to the developers to decide the type and density of houses they wanted to build.

They made that decision in function of 'the market': They built what they thought the market needed at that time. If that was large four bedroom family houses with big gardens, densities would be low. If it was one bedroom starter houses, densities would be high. For that reason the densities prescribed in the housing strategy were not adhered to.

These kinds of changes and accompanying re-negotiations during the process can be looked at from two different angles. From one angle, they can be seen as a problem caused by the incapability of the local planning authority to control the development. The house builders go their own way and the planning authority just has to keep following them and steering where possible to achieve some of its goals; the house builders are then seen as a party for whom the planning authority represents an annoying hurdle that has to be taken. From the other angle, adaptations like the ones discussed above, can be seen as an example of the flexibility of the planning system, which can respond to changes in the market. These changes will happen anyway, so the best thing is to try and decide each time on the best way to react.

In Bonn Ippendorf, there was no public decision about the number of dwellings to be realised. The parcellation was drawn up in function of the land that the owners brought in. Moreover, the number of plots did not equal the number of dwellings, because on several plots more than one dwelling was realised. The density was more a result of design prescriptions in the local plan. This plan gave maximum building volumes and outlines. Thus, it indirectly set limits for the number of dwellings to be realised on each plot. After that, the density that was realised depended on the landowners. Some of them chose to build a house for themselves, others used the plot to build several dwellings to let or sell, or sold their plot to another house builder to do so. Cases where landowners built a dwelling for themselves in combination with a dwelling to let or to sell also occurred. The density that resulted from this was not based on an overall appreciation of either design or financial considerations.

Urban design

Decisions that influence the design of the housing scheme are taken in almost all phases of the housing development process. In all cases, it started with the local plan in which the planning authority gave broad outlines for the development, such as a road structure and a general partition of the scheme into sections, often coupled with a phasing of the development. In a next step, this broad outline was worked out in more detail. Who did this and how this was done differed throughout the cases. Sometimes the municipality itself drew up the detailed plans. In other cases this was done in cooperation with the private or public-private developers. In the two cases in the United Kingdom, no detailed plan was drawn up for the whole scheme. The private developers worked out the details of the plan only when they submitted a planning application for a specific part of the scheme. three cases which give a good illustration of how the urban design was influenced during the development process are described below. First Bois-Guillaume Portes de la Forêt, where the design standards were drawn up in cooperation between the municipality and the private developer. Then Zwolle Oldenelerbroek, where the municipality determined almost entirely on its own what the scheme would look like. Finally Cramlington North-East Sector, as an example where the design of the area was to a large extent the responsibility of the house builders, although the local planning authority tried to influence it.

The development in Bois-Guillaume Portes de la Forêt started with a global plan by the municipality which indicated no more than the limits of the area to be developed with housing and the main transport axes. This was the basis for a contest in which two private developers drew up a detailed plan for the housing scheme. The plan that the municipality chose out of these two corresponded to its wish to accommodate a range of types of dwellings in a nicely landscaped, rather spacious housing scheme. Moreover, the plan had some special features that made it win the contest. Two of these have been mentioned before, a system for on-site rain water storage incorporated in a green space, and a hierarchical road system. But regarding the design of the area, the plan went further. It made a conscious effort to maintain existing hedges and trees, and it contained a planting scheme, whereby the developer would plant trees and hedges along the roads and paths. Further, design guidance regarding the houses was part of the plan. This concerned roof toppings, colours of walls, and choice of materials to be used. The objective was to realise a housing scheme that, although modern, referred in its characteristics to the traditional urban design in the area.

Following the choice of a developer and his plan for the area, the municipality created a *Zone d'Aménagement Concerté* (ZAC). This procedure provided a framework for further interactions. Although the design guidance and other design aspects were presented by the developer in the initial plan, they were adjusted to the ideas of the municipality during these interactions. The municipality had deliberately chosen a private developer to make the plan, because it felt that its own organisation was too small to have the necessary know-how and technical knowledge. The way this worked out gives a good example of how ideas can shape decision making. By producing a plan for the housing scheme, the developer initiated the decision making process about the design of the area. In drawing up this plan, the developer had in mind the general outline of the plan given by the municipality, but this was adapted to the developer's view of design quality. For example, a curve was made in the main through road which the municipality had presented as a straight road. The idea of the on-site rain water storage was completely new. And it was this idea that persuaded the municipality to select this plan. In the interactions, this was adapted and moulded into a form which both parties approved.

In Zwolle Oldenelerbroek, the municipality itself was responsible for the land development. It not only owned the land in the plan area, but also had the necessary know-how to realise the land conversion. To reach the desired level of design quality, the municipality proceeded in several stages. First, a global plan for the whole area of Zwolle Zuid (a large urban expansion of which Oldenelerbroek is a part) was drawn up. When the time came to develop Oldenelerbroek, this global plan was worked out in some more detail by the municipality in what was called a 'globaal eindplan' (global end-state plan). The plan fixed number and type of houses, and what was called 'the main structure' of the plan, a rough parcellation around the main infrastructure network. On the basis of the *globaal eindplan*, the municipality developed a '*beeldplan*', a vision of the future design and layout of the area, within the limits fixed by the *globaal eindplan*. This included for example the location and form of the edges of the development, a footpath and cycleway network, the through roads. In a next stage, the municipality specified a detailed parcellation, with the complete road layout and parking spaces, in a fully developed urban design for the development of the area.

Thus, the design of the area was specified in detail, but not fixed once and for all in the *beeldplan*. There was, however, very little space for the house builders' influence, this was limited to the design of the houses. And because they could make this choice, they also had some influence on the parcellation. Because the municipality did not know exactly what type of houses the developers would choose to build, it based the parcellation on a standard housetype and plot size. The house builders could vary a little with this, but they had to respect the road layout and the other principles laid down in the *beeldplan*. Further refinements of the *beeldplan* were made by the house builders together with the municipality in 'project teams' per section of some 200 houses of the total 1100 in the scheme. In these project teams, negotiatons about the final form of the development took place. In these teams the different departments of the municipality concerned with the housing development were represented. Each of these had its specific wishes and demands. These were adapted to the wishes and demands of the developer, so that an acceptable form of the development for all parties was found. During these negotiations, the conditions of the *globaal eindplan* (as to main structure, number of houses, densities, etc.) had to be adhered to.

A similar process of decision making about the design took place in Arnhem Rijkerswoerd. Although the developers were allowed to detail the *bestemmingsplan* further, this activity was not without supervision by the municipality. To create unity in the housing scheme, the municipality issued 'richtlijnen voor de stedebouwkundige uitwerking' (design guidelines) in the form of a booklet with rules, with which the developer had to comply in his design. These rules concerned for example the use of colours, or the building height. They came in addition to the usual design guidance that indicated things such as road width, back to back distances, and maximum distances to public spaces, which were all part of the *bestemmingsplan*. These design guidelines were concerned with the overall 'aesthetic' impact of the area.

When the development of Cramlington North-East Sector started, the municipality had presented the outline of the development in the local plan. But the details of the design were filled in only when an application for a building permission for a particular section was submitted. These applications typically concerned some 50 to 60 houses. The private house builders filled in these sections with standard house types from their collection, varying the finishings, such as window frames and roof types. This way of developing the area section by section with standard house types had important consequences for the overall design of the area. It led to a parcellation that is characterised by a main through road which feeds a number of cul-de-sacs of various length, each serving some 50 houses. When a house builder had finished the development of such a section, a next cul-de-sac was attached to the through road and a new section was developed. Groups of different housetypes appear in one cul-the-sac. But when a bigger part of the scheme is taken into account the same housetypes reccur. This type of development made it difficult to fit cycleway and footpath networks into the scheme since it was based on cul-the-sacs. Another result that was regularly mentioned was that the development did not get any 'character'. However, this seemed to be more a concern of the planning authority than of the house builders. And the latter justified this way of proceeding by arguing that each time they tried to do something 'different', they had difficulties selling the houses. Buyers apparently did not like experimentation and felt most comfortable when buying a house with a 'low profile' design.

This is not the place to discuss what is good design and what is not, and how this can be judged. Nevertheless, Blyth Valley District Council felt that it had a responsibility for the design of the area, and at some stage in the process it was concerned about the design of the development. The way in which it had an influence on the design was through the application for a planning permission. To develop a section, the house builders submitted a detailed plan of that section, which was judged by the local planning authority. If the proposed plan did not fit in with the requirements of the municipality, building permission could be refused. The grounds on which the application was judged were presented in the local (Comprehensive Development Area or C.D.A.) plan, and the planning agreement signed in 1974 or its successor signed in 1994. These documents did not allow the planning authority to exercise influence on the design of the area, provided the plans of the house builders complied with social and safety standards, such as road standards and the building code. The only way open to the planning authority was to discuss the proposed plans with the house builders, and to try and persuade them of certain design aspects.

Apart from this, the planning authority introduced the special design sites in the 1994 housing strategy. This gave them more possibilities to influence the house builders. In section 5.2, we already described how these were not very successful because on the one hand the developers were not very happy with them. But on the other hand, the residents objected also. The solution which the planning authority found was to designate 'mixed' sites, so that a developer who wanted to build traditional family houses also had to take care of some special design or affordable houses.

Public facilities

The notion of public facilities is used here to indicate all facilities that are part of a housing scheme, but which do not primarily have an aim that can easily be expressed in commercial terms. Thus it covers schools, parks, libraries, community halls, etc. The distinction between primary and secondary services is relevant here. Public facilities can be primary services, for example public spaces such as children's play spaces and small green spaces. These are usually considered to be part of the housing scheme and are therefore logically financed by the developer. In this section, the focus is on the public facilities that fall in the category of secondary services, because it is mainly here that the question of finance poses arises. It often gets explicit attention in the development process, as opposed to the issues of density and design which are not often discussed in financial terms. Below, we analyse how these discussions led to decisions about public facilities.

Bishop's Cleeve offers a good example of how an initial decison by the municipality was adapted in negotiations with other actors, in this case the house builders. Financial aspects played an important role in this. The decision making process started with the planning agreement signed in 1987 between the house builders and Tewkesbury Borough Council. There was not much open space in Bishop's Cleeve, so the Parish Council and the Borough Council wanted to use the new housing development to create some more open space. They asked the house builders to provide 5.25 hectares of open space within the housing development of Bishop's Cleeve. The house builders at first did not want to do this, but in an appeal, the inspector said that the demands of the planning authorities were legitimate and they had to provide this open space. This meant that the developers would lose 5.25 hectares of land where they culd have developed houses. Apparently, the developers preferred to offer a bigger area of land, plus a substantial amount of money, instead of 'losing' this land for housing development. Therefore they put forward 13 hectares of land they owned. This land was not in the new development, but just outside it in the green belt around Cheltenham. For that reason, the chance that it would ever be developed as housing land was very small, so the land was less interesting to the house builders than the 5.25 hectares that were within the housing area. Because of this, the planning authorities even managed to obtain an amount of 770,000 Euro from the house builders for the development of the area as a green space. This was unusual, because the notion of planning gain did not really exist at the time. The Borough Council felt that recouping some of the development gain made by the developers in this way was quite an achievement. The way it worked out resulted in the community as a whole gaining a large area of open space, but the new housing scheme remained without a large open space.

The negotiations between house builders and planning authorities did not always work out so well for the planning authorities. An example of how it could go wrong was the school site at Bishop's Cleeve. In the initial plans for the area, there was a site reserved for a primary school. But the area that was reserved appeared to be too small for the realisation of a school – partly due to the increase in population resulting from the increase in housing density – so the planning authority claimed more space. The house builders were not prepared to provide free of charge the two acres of extra land which the planning authorities claimed. According to the house builders, since this was not in the agreements, the planning authority had to pay for the land. However, the latter did not have enough money, so until 1997 when the data for this study was gathered, the housing scheme remained without a school, about which the planners at the district council felt very disappointed.

In Rennes la Poterie too, there were problems concerning the realisation of a primary school. At the start of the development process, two primary schools were planned in the area. But in the course of the development, the municipality thought that it would be better to realise only one school. The reason for this decision was that the numbers of pupils were dropping at the schools in neighbouring areas, so that there were free places there. The municipal decision to realise only one school in the Poterie meant that a certain shortage of school places had to be accepted, because in the beginning, a large proportion of the population was expected to be young couples with school-going children. In the course of five to ten years, this proportion was expected to drop as a result of the ageing of the population. However, in the case of the Poterie, the municipality was especially modest in the realisation of schools. This was the source for a lot of complaints from the inhabitants towards the municipality. A residents' association was created that played an important role in this struggle of the inhabitants against the municipality. The municipality, however, stuck to this one school. This meant that for some years, some of the children from the Poterie had to go to schools in neighbouring areas. The argument of the municipality was that there were enough places there, and that building another school in the Poterie might have been sensible at the time, but this school would be unoccupied within ten years, which would have cost a lot of money to the community.

The case of Bois-Guillaume is a good illustration of the possibilities that a low price of the land creates for expenditure on public facilities. The same mechanism can be seen at work in the case of Rennes la Poterie. In the cases of Zwolle Oldenelerbroek and Arnhem Rijkerswoerd too, the moderate prices of the unserviced building land must have played a rather similar role. But in these cases it was more concealed, because the calculations were made internally, within the municipalities. In all these cases, the municipalities had pursued an active policy of land acquisition. They had – sometimes with the help of pre-emption rights and with the threat of a possible expropriation – bought the land in advance of development at a price close to the agricultural value. The idea behind this was that in this way, almost the entire value increase caused by the land conversion could be used in the development process. In other words, the municipality prevented the first landowners from capturing part of the financial margin and in that way saved it for a later phase in the process.

Thus, the municipality of Bois-Guillaume had acquired the land in the area for a price close to the agricultural value. It was willing to sell it at the same price to a developer if the latter used the financial margin that this created to spend more on the residential environment. In practice, this meant that the developer had to provide some public facilities in the area in exchange for the right to develop it. Thus, the developer contributed to a school and a nursery on the site, and to the extension of a secondary school, in proportion to the number of new pupils the new housing scheme was expected to accomodate. The developer was responsible for providing the area with public spaces, such as a central square, play spaces and green spaces, and for equipping them. The developer also agreed to contribute in a next phase of the development to a fly over that would bridge the future bypass around the housing scheme to give the inhabitants of Bois-Guillaume easy access to the surrounding woodlands.

In Bonn Ippendorf and Stuttgart Hausen-Fasanengarten, the first landowners remained the owners of the land throughout the process of land conversion. This had an influence on the way in which decisions about public facilities were taken. In the case of Bonn Ippendorf, the only public facilities that were realised were primary services. The landowners contributed to these by providing the necessary land free of charge to the municipality. The realisation of the facilities (green and play spaces in this case) was the responsibility of the municipality. The municipality used its public powers in the procedure of the *amtliche Umlegung* to oblige the landowners to give a certain percentage of their land to the municipality free of charge. As the facilities here were primary services, which were considered to be part of the land development, up to 90% of the costs could be recouped afterwards from the landowners by means of the *Erschliessungsbeiträge* (contribution towards land development, see appendix). Because of the small size of the housing scheme and the detailed public regulations, there was not much interaction about the public facilities.

The housing scheme in Stuttgart Hausen-Fasanengarten was bigger than that in Bonn Ippendorf and included public facilities, such as a school and a kindergarten. Here, the municipality did not use its public powers in the procedure of the *amtliche Umlegung*. To make the landowners – who were also the developers – contribute to the residential environment, the municipality and the landowners agreed to realise the scheme as a *Umlegung mit freiwillig vereinbarten Konditionen*. In the negotiations about the conditions under which the municipality would designate the land for housing development, decisions about the financing of public facilities were taken. Just as in the *amtliche Umlegung*, the landowners provided the land for infrastructure and primary services free of charge. This amounted to 30% of the land. Above that, the landowners agreed to provide another 20% of their land at a reduced price. This land was to be used for social housing and public facilities, and thus formed a contribution by the landowners towards these items. It would have been much more expensive – and for that reason probably not possible – if the municipality had had to pay the market price for the land on which the facilities were realised.

Mix social/market sector dwellings

The income from the realisation of social sector housing is lower than that generated by the realisation of market sector dwellings. Instead of the social sector houses, market sector houses could have been built which would have yielded a higher income. We consider this loss of income here as expenditure on the residential environment. The case studies in the United Kingdom illustrate this. In these cases, private house builders realised both the land and the housing development. Realising social housing was not part of their objectives. Since they were not obliged (e.g. by public law) to realise social housing, they did not do it. The planning officers at the local planning authorities generally stated that they would like to see social housing realised in the schemes, but they had no means to make the house builders comply. The decision about the type of housing was taken by the developers. Apparently, when these were not forced to do so by public powers, or persuaded by financial compensations, they did not spend money on social housing.

In the cases where social housing was realised, this decision was always initiated by a public actor. In the cases of Arnhem Rijkerswoerd and Zwolle Oldenelerbroek, the number of social houses to be built was partly decided in a complicated interplay of municipal, provincial, and central government. On the basis of the calculated need for housing and the available land for housing, these parties drew up a *woningbouwprogramma* (programme for housing construction). The results of this were expressed in the *bestemmingsplan*, where the municipality told the developers what type of houses, and how many of each type to develop, on the basis of the *woningbouwprogramma*. This programme was based on the one hand on the expected need for housing in the whole province of Gelderland, calculated on the basis of demographic trends. On the other hand, it was based on the financial calculations made by the municipality. The costs of land development had to be covered by the revenues from the disposal of building plots, so the finance department in the municipality calculated the number of plots in the different housing categories that needed to be disposed of to cover the costs of development. This was important because the social housing plots yielded less money than the plots for the market sector. Because the municipality sold all the plots, it could use the higher revenues from market sector plots to finance social sector plots. This practice of 'cross-subsidising' was common practice in the Netherlands, and can be found in both cases, Zwolle Oldenelerbroek and Arnhem Rijkerswoerd.

There is also a difference between the decision making process about social housing in the two Dutch cases. At the beginning of the 1990s, central government subsidies towards the housing corporations stopped. The housing corporations have since then had to take care of their own budget. This resulted in an important change in the way the corporations work. Before – this was the case during the development of Arnhem Rijkerswoerd – if the municipality asked housing corporations to build somewhere, they usually agreed to do it. If they were then unable to let the houses and thus incurred deficits in their budget, central government would fill these with subsidies. These subsidies do not exist anymore, which means that the corporations have had to adapt their supply of houses to the demand. The corporations have started to act more like market-driven parties. Since 1996, when the central government subsidy disappeared completely, it has been left almost entirely to the corporations to build social housing without losing money on it. This has meant at the same time that the corporations have had to take a considerable risk. For that reason, they feel that the close supervision by the municipal government of their activities is not always reasonable: Since they take the risks, they want to decide when and what to build. In Oldenelerbroek, this change in attitude started to appear. The realisation of some market sector houses by a

230

corporation illustrates this. However, it did not influence the behaviour of either the corporations or the municipality very much.

Cross-subsidising of social sector plots was used in Rennes la Poterie also. The price at which the building plots were sold in this case show how market sector plots were sold at a price above the average plot price required to balance the financial account. This enabled the developer (in this case the SEMAEB) to sell social sector plots below the average price, thus cross-subsidising social sector housing. The price was expressed in the price per square metre of living space (*surface habitable*). These prices were reviewed every year. The following four categories were distinguished, with the price per square metre of living space for which the land was sold in 1998. Social rental sector (*Programme de Location Aidée* or *PLA*) at 135 Euro, very social sale sector (*accession très sociale*) at 165 Euro, social sale sector (*accession sociale*) at 217.50 Euro and free sector between 270 and 300 Euro. The land for public services was sold to the municipality at a price equivalent to that of the very social sale sector (165 Euro). Together, this resulted in an average price of around 225 Euro, which was the *prix d'équilibre* (balancing price) of the operation.

In Germany, it was up to the municipalities to decide whether or not they want to realise social housing within their boundaries. The municipality of Stuttgart initially wanted to develop the whole area of Hausen-Fasanengarten with social housing. To this aim, the municipality referred to a paragraph in the *Baugesetzbuch* (the building code) which facilitated the designation of areas for 'besonderen Wohnbedarf' (special housing needs). This meant that many of the landowners would not have been able to develop their own land, because the social housing was to be for persons within certain revenue limits only. Moreover, if they wanted to sell their land, they would only have received a low price, because the construction of profitable free sector dwellings would not have been possible on the site. Complaints were raised against this, and they were sustained: In 1992, the designation of the whole area for social housing was found to be unlawful by the *Bundesverwaltungsgericht* (federal administrative court) in Mannheim.

There were two reasons why it was judged unlawful. The first one was that – according to the *Bundesverwaltungsgericht* – the term *besonderer Wohnbedarf* had not been interpreted correctly by the municipality of

Stuttgart. In the building code, this referred to disabled people for example, or people who have special needs for other reasons, but not to people with low income for whom social housing programmes were intended. The second reason was that in any case, it was not legal to designate an entire plan area for *besonderen Wohnbedarf*. This can be applied only to parts of the plan area. As a result of this judgement, a new *Bebauungsplan* for the area had to be drawn up. The changes in this new plan concerned not so much the design of the area, but the prescriptions about social housing. This was changed into a prescription that houses in the area had to be affordable for lower or middle income groups. This prescription implied that a certain price level should not be exceeded. Moreover, for parts of the plan the municipality maintained the prescription of *besonderer Wohnbedarf*.

The municipal council still wanted to realise social housing. However, this was very difficult due to the high land prices in the area. Therefore, a means had to be found to moderate the price of at least part of the building land. In April 1993, a *Städtebauliche Entwicklungsmaßnahme* (urban development measure) was proposed by the planning department. This instrument enables the municipality in a specified area to (compulsorily) purchase the land required for development, for a price that does not include hope value. That means the land is acquired for a relatively modest price, which enables the construction of less profitable uses on the land, such as social housing. However, it is a 'heavy' instrument. Indeed, it is the most drastic measure which a German municipality has to realise urban development. For that reason, it can only be used when '...the public interest' requires consistent preparation and speedy execution...' (David, 1995: 165). On second thoughts, the municipality of Stuttgart did not try to justify this general interest. It decided that the *städtebauliche Entwicklungsmaßnahme* was too drastic an instrument, and it did not want to use it: It would have interfered too much with the property rights of the separate landowners.

Moreover, before a municipality decides to carry out a *städtebauliche Entwicklungsmaßnahme*, it has the statutory obligation to investigate whether the development might be carried out using another – less drastic – instrument. If possible, the use of a *städtebauliche Entwicklungsmaßnahme* should be avoided. This investigation led in Stuttgart to the choice to develop the area by the use of an *Umlegung mit freiwillig vereinbarten Konditionen* (urban reparcellation on a voluntary basis). As to the extent to which it interfered with the property rights of the landowners, this could be considered as a measure between *amtliche Umlegung* and *städtebauliche Entwicklungsmaßnahme*. So the municipal policy for the realisation of social housing had an important influence on the type of development process that was chosen in this case. It allowed the municipality to claim 20% of the land from the landowners at a price of 90 Euro per square metre, which allowed the realisation of social housing.

5.4 The development process and the residential environment

In this chapter, the decisions that were taken during the different stages in the process of housing development have been made explicit. We have explained the decisions concerning the residential environment as a result of both financial and institutional considerations. At the end of this chapter, we resume the headlines of the analysis carried out above. In both the financial and the institutional analyses as they were carried out in chapters three and four, we have distinguished stages in the development process. In general, these stages correspond both to phases when a financial margin can appear (see table 3.1) and to the activities around which the institutional analysis was structured (see table 4.2). The first two activities in the process, i.e. 'identification of development opportunities' and 'land assembly', correspond to only one phase where a financial margin can appear. During the identification of development opportunities, nothing is traded. The land remains in the hands of the first landowners. Only when the activity of land assembly starts, do developers start to buy land. A financial margin on the market for unserviced building land occurs then. For that reason, the first two activities in the development process are taken together here.

–   *Identification of development opportunities and land assembly: financial margin (on the market for unserviced building land) received by first landowners*

233

In this stage decisions about the residential environment, for example concerning housing density and the mix of social sector and market sector dwellings, were taken in the development plan (except in Cramlington and Bishop's Cleeve). However, these decisions were not yet realised in the field. A possible margin went to the first landowners and was not used directly for expenditure on the residential environment. When the plans for expenditure on the residential environment influenced the residual calculation of the developers buying the land, the financial margin for the first landowners was reduced. This partly explains the low price of land acquisition in Arnhem Rijkerswoerd and Zwolle Oldenelerbroek. The use of pre-emption rights (Rennes la Poterie) and the threat of expropriation (Rennes la Poterie, Bois-Guillaume Portes de la Forêt, Arnhem Rijkerswoerd, Zwolle Oldenelerbroek) also reduced the price of land, hence the financial margin for the first landowners. This allowed a financial margin to be used in another stage for expenditure on the residential environment.

–   *Land development: financial margin (on the market for serviced building land) received by land developer*

In this stage, the basic structure of the housing scheme was realised. Decisions taken previously in the development plans were realised 'on the ground' and used (and adapted) in interactions between the planning authority and the developers. As a result, this stage was important for the residential environment. The road structure and the parcellation realised in this stage clearly set the outlines for the urban design. But also the housing density, the mix of social sector and market sector dwellings and the public facilities were closely related to the parcellation. The structure of the housing scheme that was laid down in this stage set conditions for what could be realised. In most of the cases, tools or procedures aimed at spending part of the financial margin in this stage on the residential environment could be observed. The procedure of the *Zone d'Aménagement Concerté* in Bois-Guillaume Portes de la Forêt and Rennes la Poterie and the active policy of land acquisition and land development pursued by the municipalities in Arnhem Rijkerswoerd and Zwolle Oldenelerbroek were examples of this.

234

- *Housing development: financial margin (on the market for new housing) received by house builders*

The activities in this stage had a great impact on the appearance of the housing scheme. The urban design was largely determined here since the overall structure of the development that had been laid down in the preceding stage was now filled in with houses and other buildings. The design of the buildings determined to a large extent the appearance of the whole scheme. In some cases, the financial margin received in the first instance by the house builders was used for expenditure on the residential environment. In the cases of Cramlington North-East Sector and Bishop's Cleeve, the same private parties that developed the land also built the houses. In the planning gain agreements signed between these private house builders and the local planning authority, the private house builders agreed to spend part of the financial margin on the residential environment. In Bonn Ippendorf, the house builders were made to contribute to the land development and hence to the residential environment by means of the *Erschliessungsbeiträge*. Decisions about what this contribution was spent on were made by the planning authority, who also carried out the land development.

- *Final ownership/use: financial margin (on the market for second-hand housing) received by final users*

In this phase, the actual development of the housing scheme was finished. The only aspects of the residential environment that played a role in this phase were facilities (commercial and public) that chose to settle there. This sometimes led to problems as in Rennes la Poterie and Bishop's Cleeve where for different reasons, there were problems to get a primary school to settle in the scheme. In none of the cases were arrangements made to redirect a possible financial margin on the market for second-hand housing into the residential environment.

235

6 Land policy and housing development: lessons to be learnt

6.1 Recapitulation

To introduce the conclusions from the research reported in the previous chapters, this section provides a short recapitulation of the argument that has been developed.

The issues

The objective of the research reported in this book is to investigate how during the process of housing development decisions are taken that influence the residential environment. There are several reasons why this study was undertaken. In the first place, there was the intellectual curiosity about how public policy and market mechanisms work together in the production of housing. This somewhat academic question suddenly became acute because of the changing situation in the Dutch land market since the beginning of the 1990s. Rising prices and a sudden active involvement of private parties in this market resulted in the municipalities not being able to continue their active policy of land acquisition. Private developers, who until then had not been active on the land market, started to buy large pieces of land in areas designated for development. This changed the position of the municipalities towards private developers, but equally towards other parties in the process, such as housing corporations and first landowners. This can be placed in a wider context of government withdrawal from development processes and an increasing confidence in the regulating principles of the market, which can be observed throughout Europe.

The study aims to improve the understanding of the (housing) development process on two levels. On a theoretical level, this contribution lies in the combination of insights derived from a micro-economic analysis and an institutional policy analysis. The study of a variety of development processes in different countries allows us to formulate theoretical insights that can be applied to a wide range of development processes. We return to this aspect of the study in section 6.2. We feel that some of the insights formulated in this study have a broader field of application than the housing development process. We want to place our ideas of how actors in development processes influence the behaviour of others, and how in turn this influences the outcomes of these processes, within the theory about institutional policy analysis. In section 6.5, the outcomes of this study are therefore used to open a broader discussion about 'institutional capacity building' (see Healey, 1997) and about 'enrichment' (Teisman, 1992) in planning processes.

At a more practical level, the study investigates how changes in the development process might change its outcomes. The broad analysis carried out in different countries – hence in different social, economic, and institutional contexts – allows the book to shed new light on questions, such as how public facilities or social housing can be financed, how good ideas concerning urban design can be turned into reality, or in a more general sense, 'how they do it over there'. This must not be interpreted in a deterministic way. What the study does, is to reconstruct retrospectively how in eight development processes the final outcomes came about. Thus, it provides actors in development processes with a framework to reflect upon their actions, to interpret what happens during a development process. The 'substitute experiences' of the case studies are used to develop ideas as to how different forms of the development process might lead to different outcomes. In section 6.3, this aspect of the study is dealt with. On the basis of this, in section 6.4, we take up the gauntlet and try to formulate policy options to deal with the Dutch situation, which was one of the reasons to carry out this study.

The core of the study are the eight case studies of housing development processes that were carried out in four different countries (two in each country). Although this gives the study a cross-national character, it is not the aim to compare land policy and housing development on the level of the countries involved. Others have done that before (e.g. the European Urban Land and Property Market series, 1993, 1994; Barlow and Duncan, 1994; Newman and Thornley, 1996; Balchin, 1996). For more details about land policy and housing development in the studied countries, we gladly refer to these volumes. The subject of this study are the separate housing development processes. The reason for carrying out the study cross-nationally is to allow a broader range of housing development processes to be studied. The development process of the eight housing schemes – which starts with the identification of development opportunities and terminates with the sale of the completed houses and the handing over of public spaces – has been studied with the use of an analytical framework combining insights derived from micro-economic analysis and from institutional policy analysis.

The choice for a case study approach has been made because this allows us to translate the rules and practices of public intervention on the urban land and property markets on the national level into what actually happens 'on the ground', on the local level. It has consequences for the research methods used to gather the data and for the type and scope of conclusions that can be drawn. As in any case study research, different methods of data gathering are combined, allowing the researcher to approach the studied phenomenon from different angles. Data about the cases of housing development processes have been gathered by means of interviews with different participants, by the study of official documents and maps, by other printed sources, such as newspaper articles, and by site visits. Thus, data reflecting a whole range of reconstructions of (parts of) the processes became available. These have been analysed and combined into this study's reconstruction of the process. The theoretical notion that is used to describe this activity is double hermeneutics: The researcher gives his interpretation of the interpretation that others have made of the development process. The aim of this exercise is to provide people involved in land policy and housing development with a useful perspective to interpret and evaluate the events and interactions in a development process, and to develop ideas about how changes in the process might lead to changes in the outcomes.

The results

The results of the investigation are presented in the following sections of this chapter. However, we want to introduce them briefly here by recalling two central assumptions that have been used in the study, and by presenting what the investigation of these assumptions has led to. These assumptions are: the central place of the process of land conversion in the housing development process, and the influence of a possible financial margin when housing is developed on expenditure on the residential environment.

In section 1.6, we argued why we assume that the process of land conversion, i.e. the land assembly and land development, occupies a central place in the development process of housing. More specifically, we assumed that the actor who (temporarily) owned the land during the land development would play a key role in the development process. This central assumption guided the choice of the cases. The question poses itself: Was it justified to base the choice of the cases on this variable? Did processes with different (temporary) landowners during the land development have different characteristics? The answer to this question has two sides. On one side, it has become clear that the owner of the land during the land development plays an influential role in the process. Building land is a scarce resource and therefore the actors owning it have a certain power: If they do not cooperate, then nothing can be done. However, generally, this power is acknowledged and compensated for by regulation. That is the other side of the answer. The differences in the outcomes of the process when the temporary landowners were different were not so big.

In the terms of our study, this can be explained as follows. Resources – building land in this case – are only one source of power. We found that when an actor does not have this power, he might compensate this by another type of power, i.e. rules or ideas. This sounds easier than it is, because often the actors in the process cannot create new rules which they themselves can impose upon the others. This is especially true for the private actors. Changing the rules would involve interference from higher levels of government, and in any case would not alter the position of the private actors since they cannot impose rules upon others in the same way the public actors can by using public law. What remains are the ideas by which the behaviour of the others can be influenced. This option is open to all actors. However, it is difficult to put into practice and only in a few cases did ideas actually alter the course of the development process. In section 6.3, when we investigate how changes in the process might change the outcomes, we return to the possibilities for tinkering with the power balances and thus for influencing the process.

In our investigation, the power balances between the actors were important because they determined the influence of each of the actors on the use to which the money generated in the process was put. To investigate this, we introduced the notion of the financial margin (the second assumption). In chapter three, we have shown whether and where such a margin appeared in the cases. The use of the notion turned out to be helpful in the financial analysis of the development processes. The link with the decisions about the residential environment was harder to trace. To make it, we have to stress that the financial margin as used here is only an analytical tool. In the financial calculations in section 3.2, we have shown that in some cases it appears explicitly. However, the figures presented in that section do not always correspond with the importance of the financial margin for the residential environment.

The actors in the process were often quite aware of who received a possible financial margin. This induced them for example to buy unserviced building land early, before it had acquired any 'hope value' (in the cases of Zwolle Oldenelerbroek, Arnhem Rijkerswoerd, Rennes La Poterie, and Bois-Guillaume Portes de la Forêt). In other cases it induced the actors to create special procedures to capture a value increase caused by the development (Bonn Ippendorf, Stuttgart Hausen-Fasanengarten), or to impose contributions towards secondary services – in the form of planning gain agreements – on the developers (Bishop's Cleeve, Cramlington North-East Sector). The first two ways of acting mentioned above were aimed at not letting the financial margin 'leak' out of the process towards the first landowners. In the cases of Bishop's Cleeve and Cramlington, this could not be prevented because the land had already been acquired by the developers. Therefore the aim was to make the developers spend part of the financial margin on the residential environment instead of keeping it as a profit. So although the financial margin is sometimes hard to distinguish, it offers a useful tool to describe why actors acted in the way they did concerning the expenditure on the residential environment. We use it again in section 6.3 to describe what changes in the process could be devised to influence the expenditure on the residential environment.

6.2   The value of the analytical framework

As argued in chapter two, the perspective on housing development that has been worked out in this study is only one perspective among a number of others that exist or can be developed. Each perspective would yield other, complementary insights. In this section, we focus on the specific contribution towards the understanding of housing development processes that this study provides. A part of this specific contribution lies in the cross-national perspective that is adopted. This is dealt with first, after which we turn to what could be called the content of the framework.

The interest of a cross-national perspective

In chapter two, we have argued why and how this study was carried out cross-nationally. The main argument was that this allowed a broader range of types of housing development processes to be studied. As a result, the scope of the approach could be increased. Moreover, by applying our approach to housing development processes in different countries, we apply it in different economic, social, and institutional contexts. If it can be used in these different circumstances to increase our understanding of housing development processes, then that is an indication of the robustness of the framework. Are the difficulties and uncertainties inherent in the choice of a cross-national perspective compensated by these – or maybe by other – advantages? This question is dealt with here.

An important conclusion is that it is possible to study housing development processes at a local level in different countries using the same analytical framework. The same or similar kinds of activities, actors, roles, and interactions occur in each of the studied countries. Of course, there are a many differences between the countries. Social and cultural differences are undoubtedly important for the final appearance of the housing scheme. In Germany, the dominant culture as regards housing development is one of individual people building their own house in a later stage of their life and then remaining there, after having rented a house for a long period. In the United Kingdom, people tend to buy houses much earlier in their life, and they envisage changing house a number of times in the pursuit of their 'residential career'. The houses they buy are usually produced in large developments by house building companies. This leads to very different forms of residential development and hence of residential environments in these two countries.

However, this is not the primary concern of this study. This study's focus is on how decisions about (expenditure on) the residential environment are made. These decisions are influenced by social and cultural differences between the countries in two ways. In the first place, the institutional structures (government structures, structure of the building industry, etc.) within which housing development takes place are shaped by the social and cultural context in a country. Hence the institutional structures differ. In the second place, the preferences of the people who make the decisions differ according to their cultural background. This study's aim is not to describe how social and cultural differences lead to other institutional structures and other preferences concerning the residential environment. The institutional structures and culturally determined preferences in each country are taken as a starting point. It is taken as a fact that they influence the form of the development process and through that the power balances, and the occurrence and division of financial margins. We study how these power balances and financial margins influence decisions about the residential environment. We are not so much interested in studying whether a possible financial margin is spent on, for example, public facilities or on social housing. What interests us are the mechanisms behind decisions that influence the residential environment, not so much the resulting residential environment itself.

When considered in this way, the differences between the countries appear like different ways to deal with the same kind of questions. Examples of these questions are for the public actors: How to provide sufficient housing to suit the demand (or the need)? How to create a residential environment that corresponds with the wishes and demands of the inhabitants? How to make sure that social and safety standards are adhered to? How to finance public facilities, or social housing? For the private actors the questions are: How to respond to fluctuations in the housing market? How to ensure the continued existence of my company? How to do a good job? For the landowners: How to make sure that I can continue my business elsewhere? How to get a fair price for my land?

Studying the way in which these questions are dealt with in different countries, and how decisions about the answers to them are produced, allows us to learn from solutions that others have found. When staying at the level of overall regulation, this is very difficult. The legal and administrative systems, and the planning and housing policies are very different at first sight. Studying them on the level of individual development processes allows us to see what their results are in the everyday practice of housing development. Then the differences appear to be less insurmountable, and 'how they do it over there?' becomes an interesting question that can provide real lessons for 'how we could do it'. We have judged these lessons to be more important than the difficulties with the research design and methods that this cross-national character implies, as discussed in chapter two.

Driving forces behind housing development

The process of housing development has been described as a process in which a number of actors – each with his particular role or combination of roles – carry out a number of activities. These activities result in the realisation of a housing scheme. Each of the actors in this process has his own objectives. The actors are interdependent in that they need the others to realise their objectives. For that reason, the actors enter into interactions. In these interactions, the decisions about the outcomes of the development process – and hence about the residential environment – are taken. The interdependence between the actors is not symmetrical. Therefore, although there is not a single actor in charge of the process taking all the decisions, the influence of the different actors on the outcomes of the process varies. And that is where it becomes interesting. Apparently, there are factors in the process that are unevenly distributed over the actors and that make their influence on the course of the process more or less important. Formulated like this, this is obvious. The interest of this observation is, however, that it gives us an approach to the central aim of this study, which is, finding out how changes in the process of housing development might change the outcomes. If we know why in a particular process the influence of certain actors is bigger than that of the others, we might at the same time have a key to changing this situation.

What we want to find out, then, are the 'driving forces' that make each process of housing development proceed in the way it does. On the basis of earlier studies of (housing) development processes, we have discerned two main driving forces, which are money and power (see section 1.3). Or, to put it in more academic terms, financial and institutional considerations. These two considerations enable or constrain the ability of the actors to pursue their objectives. They determine the interdependence between the actors, and are the input for the interactions. To allow a deeper understanding of the way in which these driving forces influence the course and the outcomes of the development process, we have first analysed them separately.

The analysis of the financial aspects of the housing development process is constructed around the notion of the 'financial margin'. The idea behind it is that expenditure on the residential environment should be with the money generated in the development process of the housing scheme concerned, to avoid charging tax-payers for facilities that benefit mainly people in the newly developed area. The difference between the revenues gained by the sale of the houses in a housing scheme and all the costs – including normal profits – that are incurred in the process of development is the financial margin. This limits the room for expenditure on the residential environment over and above the minimum level of quality determined by social or safety standards. In chapter three, we have demonstrated how micro economic-theory – the residual theory of land prices – allows us to distinguish such a margin in different stages of the development process. To what use a possible financial margin is put,

245

depends upon the objectives of the actor who receives it, and upon the power of the other actors to influence the use to which it is put. This is the theory. In practice, it turned out in many of the cases that it was not possible to distinguish a financial margin as such. It was not an item that occured on the financial accounts of the operations. It was deduced from the numbers that appeared on these accounts. The notion of financial margin must therefore primarily be seen as an analytical tool, and not as a recognisable variable in housing development processes.

The starting point for our institutional analysis of the housing development process was Healey's 'institutional model of the development process' (1992), which we adapted and elaborated to serve the aim of this study. The analysis proceeds through three levels. The first is a chronological overview of the events that occur in the development of a housing scheme. On the next level, this description of the process allows us to distinguish the actors in the process, the role(s) they play, and the activities they carry out. The description of the actors also includes an investigation of their objectives in the process. On the third level of analysis, the focus is on relations of power and dependence between the actors. This requires a study of the interactions between the actors, because in these interactions the power balances become visible. The strategies of the actors in the interaction evolve from a combination of their objectives, and their ability to pursue them (i.e. their power) in the interactive policy network of the housing development process. Three grounds for power are distinguished: rules, resources, and ideas. In the way in which they are applied in this study, rules refer either to public or to private law regulations that govern the relations and interactions between actors. Resources are for example land rights, labour, finance, and information. Actors who have the resources can use them to influence others. Ideas can have a binding force in the development process which allows the actor who has a good idea to use it to influence the behaviour of other actors. The reason why actors would want to use their power is to try and achieve more of their own objectives.

To combine the financial and the institutional analysis, with the aim to understand better how development processes work, we refer back to the notions of market, hierarchy, and network that are recognised by institutional economists as coordinating principles for social activities (see section 1.6). If we look back at the cases, we can observe all types of coordination, applied to various extents, in all the housing development processes studied. The dominant type of coordination in a particular process is closely linked to the power balances and hence to the outcomes of that process. In chapter four, three different bases for power of the actors in the development process have been presented, i.e. resources, rules, and ideas. Here, we put forward the idea that these bases for power can be linked to the different principles of coordination. Although we have to stress once again that in all the cases studied, elements of all three principles of coordination could be observed. It is possible, however, to use the prevailing mode of coordination in the different cases to explain the degree of influence of the different actors involved.

In a market, where price competition is the central coordinating mechanism, resources are the central basis for power. As a result, when in a housing development process the market is the prevailing mode of coordination, parties who can bring important resources, e.g. land or money, into the negotiations have much power to influence the behaviour of the others. This is most clearly illustrated in the case of Cramlington, and to a somewhat lesser extent in Bishop's Cleeve. Although counter intuitive, the cases of Arnhem Rijkerswoerd and Zwolle Oldenelerbroek were also largely coordinated by the market mechanism. In these cases the municipality owned the land during the land development. The resource of the land gave it an important power base in the interactions with the other parties. It effectively had a monopoly position on the market for building land. The municipality as a public actor could also use rules, for example the local land use plan, to exercise power over other actors. In that sense, the process had characteristics of a hierarchy. But in all cases, all coordinating principles functioned alongside each other. The case that showed most strongly characteristics of a hierarchy was Bonn Ippendorf. In a hierarchy, rules are used to coordinate the activities of the different actors involved. In Bonn Ippendorf the whole process was carried out within an administrative procedure.

The coordination in a network is a matter of trust and cooperation. In this study, such notions are put under the category of ideas. The binding force of ideas, which we have described also as communicative power, is required in a network to enable coordination between the actors involved. The case that had most of the characteristics of a network was that of Stuttgart Hausen-Fasanengarten. A mobilising idea – that of the *Umlegung mit freiwillig vereinbarten Konditionen* – and cooperation were central in the development process in this case. In Bois-Guillaume Portes de la Forêt, trust and cooperation, or in our terms ideas, equally played a very important role in the coordination between the actors.

Characterising the case of Rennes la Poterie in one of the three above-mentioned categories is even more precarious than for the cases already mentioned. In this case, rules (i.e. the procedure of the ZAD and the right of pre-emption that this created for the municipality) were used to obtain the land. The land then served as a resource, allowing the municipality (and in a next phase the public-private SEMAEB to whom the municipality had sold the land) to use the market as a coordinating principle in the realisation of the plan. In this example, the mix of coordinating principles at work in a housing development process was more apparent than in some other cases. But this mix existed in all of the cases. This has to be taken into account in the next section, where the aim is to show how changes in the process of housing development might change the outcomes.

6.3   Changing the process to change the outcomes

We want to provide insights into the decision making process about the residential environment in housing development processes. For a governmental organisation which has a responsibility for the control of development, and hence for the residential environment, such insights can be very useful. A local planning authority, or central government reflecting upon policy guidance and legislation for local government, would want to know how changes in the process of housing development might change the outcomes. In this section, we take the point of view of such a planning authority to investigate this. First, we depict the line of thought that a planning authority can follow to reflect upon possible new tools, new procedures, or new 'ways of doing things' for influencing the residential environment. At the end of this section, it is shown how this can be translated into concrete policy measures.

Strategies and negotiations in housing development processes

For the realisation of their objectives, the actors in the development process depend upon each other. Conversely, this means that each of the actors has a certain power over the others. This power is used either to realise the desired residential environment by cooperation if the objectives of the actors correspond, or by regulation if the different objectives cannot be brought into line. In the case studies, we see how this leads to interactions between the different parties. These interactions often take the form of negotiations, which can be seen as lying between regulation and cooperation. This illustrates the view of decision making in housing development processes developed in chapter five. Decisions are not made by a single actor, but by interdependent parties interacting to realise their objectives. However, the interdependence is not symmetrical. Sometimes actors can greatly influence the behaviour of others, and by doing so they can come closer to realising their own objectives.

Housing development processes can be seen as negotiating processes in which the participants have to give and take. To understand why the actors act as they do in this negotiating process, we need to know what the objectives are on which the actors base their strategy. The different participants who receive a possible financial margin on the markets for unserviced and serviced building land in our case studies were:

– the local planning authority;
– a private developer or house builder;
– a public-private partnership; or
– the original owners of the land.

In the case studies we have seen what these different actors would use a possible financial margin for. On the basis of these cases, we can summarise as follows the objectives and strategies of the differerent actors concerning the use of the financial margin. It is a simplification, but worthwhile because we found in the case studies that each separate type of actor did have particular objectives.

Local planning authorities are concerned with the residential environment, as a part of their responsibility for controlling development in general. In one of the cases, Zwolle Oldenelerbroek, a municipality received the financial margin. As it was the municipality that in this case provided the primary as well as the secondary services, it could largely shape the residential environment according to its own ideas. The municipality mainly aimed at covering the costs of production. This calculation was made in advance. The municipality estimated the income from the sale of building plots and drew up a plan in which the income was balanced with the costs of realisation. In other words, the plan of the municipality was drawn up in such a way as to reinvest a possible financial margin in the residential environment. As a result of this, it was difficult to evaluate what the financial margin in the development process of Zwolle Oldenelerbroek really was. Moreover, the objective of the municipality could be to make profits that could be used to reach other objectives that were not related to the housing development. This was the case in Zwolle Oldenelerbroek, where the positive balance on the financial accounts was used to cross-subsidise less profitable housing schemes. This was not the primary aim of the municipality, however, but more a positive spin-off from a succesful development.

The main objective of the private developers (either the house building companies in the cases of Bishop's Cleeve and Cramlington North-East Sector, or the land developer in Bois-Guillaume Portes de la Forêt) was to sell their product, and to make the necessary income to ensure their continued existence. The private developers were, in certain respects, interested in the quality of the residential environment. Three reasons were given for this interest: First, it is always rewarding to do a good job. The job of private developers is to realise housing schemes. They generally try to do that in such a way that their product is appreciated by the residents and the other parties concerned. The second reason is that if the products – the houses and their environment – are of good quality, the developer can set its prices higher. There will be a break-even point, where

250

any extra expenditure equals the increase in income that results from higher prices. A private developer tends to try to reach this point, within the limits set by local government. The third reason for private developers to be interested in spending money on the residential environment is also inspired by the necessity to ensure their own continued existence, and can be described as investing in goodwill. If they do a good job, and people (both partners in the development process and house buyers) are satisfied, the developers are more likely to be invited for the realisation of other housing schemes.

When a public-private body enjoys a financial margin, the use to which it is put depends upon what the partners have agreed. Each case will be different. In our study, the case of Rennes la Poterie gives an example of what can happen. In this case, the rules for the public-private body were very strict. There had to be two separate financial accounts, one for the public-private body, in which it registered its internal costs and revenues, and one for the operation, where the costs of land development and the income from the sale of building plots were registered. It was legally not allowed that there might be any permeability between these accounts. The public-private body earned in this case only the fee that it had agreed with the municipality. A possible financial margin in the development process was gained by the municipality. Therefore, in this case of public-private partnership, more or less the same logic applied as in the case of Zwolle Oldenelerbroek, where it was also the municipality that received a financial margin. The primary aim was to realise a housing scheme with a certain quality. The plans were therefore drawn up in such a way as to balance cost and income. If a financial margin appeared, this was welcome and could be used to cross-subsidise other housing developments.

In a number of cases, the original owners of the land received at least part of the financial margin. Two different situations occurred. In Zwolle Oldenelerbroek, Bishop's Cleeve, Cramlington North-East Sector, Bois-Guillaume Portes de la Forêt, and Rennes la Poterie, the first landowners no longer played a role in the process after having received the price of the sale, which in all these cases included a financial margin. The first landowners were not interested in spending the financial margin on the residential environment. The financial margin 'leaked' out of the process, and could not be spent on the residential environment. In the cases of Bonn

251

Ippendorf and Stuttgart Hausen-Fasanengarten, the situation was different. In these cases, the original landowners retained their land, and therefore continued to play a role in the development process. In Bonn Ippendorf, the official procedures of the *amtliche Baulandumlegung* and the *Erschliessungsbeiträge* forced the landowners to use part of the financial margin for the realisation of the municipal development plan. The case of Stuttgart Hausen-Fasanengarten showed an alternative solution, where the contribution of the landowners was arranged in negotiations in which the municipality used the local plan as 'loose change'.

Tinkering with the power balances

Above, we implicitly gave the key as to how changes in the process of housing development might change the outcomes of that process. We described the objectives which the different actors tried to reach, and we stated that the power they had to influence the other actors with the aim of realising their own objectives was divided asymmetrically among them. According to this reasoning, two ways of influencing the outcomes of the development process can be distinguished. One is to change the power balances between the actors, thus increasing or decreasing their influence on the outcomes of the process. The other is to change the division of roles between the actors. According to their role in the process, different actors with different objectives, received a possible financial margin. Changing the roles carried out by the different actors would influence who received the financial margin. If another actor, with other objectives, received a financial margin, this would have consequences for how much of this could become available for expenditure on the residential environment. In other words, if another actor received the financial margin, he would use it in a different way and this would influence the outcomes of the process. Although these two ways of influencing the outcomes of the housing development process work through different mechanisms, they both involve tinkering with the power balances. The final step in this analysis is to translate these insights into policy measures. To take that step, we first return to the case studies to describe the different ways in which planning authorities exerted influence there.

Different approaches from local government towards the expenditure on the residential environment can be distinguished. In the first place, we can distinguish on the one hand between planning authorities which influenced the expenditure on the residential environment by themselves being directly active on the market for building land. We call this direct involvement of the planning authority in the land market an "active land policy". This occurred to various degrees in the cases of Zwolle Oldenelerbroek, Arnhem Rijkerswoerd, Rennes la Poterie, and Bois Guillaume Portes de la Forêt. On the other hand, there are the cases where the planning authority was not directly active on the land market. It did not buy land land and it left the responsibility for the land development to the landowners. The role of the planning authority when it followed such a 'passive land policy' was to guide and supervise private sector development. This occurred in Cramlington North-East Sector, in Bishop's Cleeve, and in Stuttgart Hausen-Fasanengarten. Bonn Ippendorf was in an intermediate position because the planning authority did not actively purchase the land, but it did carry out the land development. Because the land was not acquired by the municipality, we consider this as a passive land policy.

In the second place, we made the distinction between coordinating principles that can be used in the development process, i.e. the market, the hierarchy, and the network. In section 6.2 we argued that in the cases of Bishop's Cleeve and Cramlington North-East Sector the market was the dominant coordinating principle. In the cases of Arnhem Rijkerswoerd and Zwolle Oldenelerbroek, the market was an important coordinating principle besides hierarchical coordination. In Bonn Ippendorf, the hierarchy was the main coordinating principle. The cases of Bois-Guillaume Portes de la Forêt and Stuttgart Hausen-Fasanengarten had many characteristics of a network as coordinating principle. We used the example of Rennes la Poterie to argue that in any case, different coordinating principles were used.

Here, we want to use this classification of different ways that can be used to exert influence in housing development processes in order to explore the policy options or tools that planning authorities can use to change the outcomes of the development process. Both distinctions made above can be combined into table 6.1. This gives us a framework for classifying the different policy measures that were used to influence expenditure on the residential environment in the cases studied. This table can serve as a basis for the choice of policy measure to apply in a particular situation, for it enables us to relate this choice to the available or preferred mode of intervention of the planning authority.

Table 6.1   **Land policy measures for housing development in the case studies**

	Market as coordinating principle - power based on resources: money, land - relations regulated by private law	Network as coordinating principle - power based on ideas: trust, cooperation - relations regulated by contracts	Hierarchy as coordinating principle - power based on rules: administrative orders - relations regulated by public law
Active land policy	- Amicable land acquisition (Zwolle Oldenelerbroek, Arnhem Rijkerswoerd, Rennes la Poterie, Bois-Guillaume Portes de la Forêt)	- *Zone d'Aménagement Concerté* (Bois-Guillaume Portes de la Forêt)	- Pre-emption right (Rennes la Poterie) - Expropriation (Rennes la Poterie, Zwolle Oldenelerbroek, Arnhem Rijkerswoerd, Bois-Guillaume Portes de la Forêt)
Passive land policy	- Public private partnership (Rennes la Poterie)	- *Freiwillige Baulandumlegung* (Stuttgart Hausen-Fasanengarten) - Planning gain agreements (Cramlington North-East Sector, Bishop's Cleeve)	- *Amtliche Baulandumlegung* (Bonn Ippendorf) - *Erschliessungsbeiträge* (Bonn Ippendorf)

The table only gives an overview of the different policy options encountered in the investigation. The strengths and weaknesses of each option depend upon the situation in which it is applied. To understand the functioning of the land policy measures described we refer to the descriptions of the cases in the preceding chapters.

Towards tools for changing the development process

One of the insights derived from this study is that a financial margin can arise at different stages of the housing development process. There is a strong argument in favour of using this financial margin for expenditure on the residential environment, i.e. the 'rational nexus' argument. It is in the interest of the planning authority, and it seems to be accepted by the public, that expenditure which benefits disproportionately the residents of a particular development, and which arises directly from that development is paid for, not by the general taxpayer, but out of the financial margin generated during the development process. All of the policy options presented in table 6.1 can be understood as being aimed at just that. To make that clear we take a closer look at the different procedures and instruments encountered in the case studies, to answer the question in which way the use of a possible financial margin is influenced. If the planning authority wants to be able to use (part of) the financial margin to spend on the residential environment, it is necessary to avoid a situation in which some of the financial margin is captured in the early stages by actors who cannot be obliged or persuaded to spend it on the residential environment. In the cases, several approaches can be distinguished for the local planning authority to pursue this aim.

–   One approach of the local planning authority to influence expenditure on the residential environment is to make clear to all the (private) actors well in advance that they might be required to contribute financially. Then they can take this into account when determining costs and prices. In several cases, tools were used that were aimed at this goal, among others. The *Zone d'Aménagement Concerté*, as it was used in Bois-Guillaume Portes de la Forêt and Rennes la Poterie, and the *amtliche Umlegung*, as in Bonn Ippendorf, created legal powers so

255

that it was clear in advance that the private actors would have to contribute to the public facilities.

- Another approach is to convince other actors in the early stages of the housing development process that it is in their interest to work closely and openly with the public authority. The procedure of the *Zone d'Aménagement Concerté* is also an example where this way was used to influence expenditure on the residential environment, like that of the *freiwillige Umlegung* which was used in Stuttgart Hausen-Fasanengarten.
- In the cases where private actors acquired hard legal rights, by buying the land as happened in Cramlington and Bishop's Cleeve, the planning authority felt that it had only few possibilities to obtain a contribution from the private actors. From this it can be concluded that it is in the interest of the local planning authorities to avoid creating legal rights for private actors before negotiations about financial contributions take place. This can be seen as a third approach to influence expenditure on the residential environment. Actively buying the land as in Arnhem Rijkerswoerd, Zwolle Oldenelerbroek, Bois-Guillaume Portes de la Forêt, and Rennes la Poterie is a way to avoid a situation in which developers have legal rights before negotiations about the development take place.

These approaches can each be pursued in ways as described above under the heading 'tinkering with the power balances', either without changing the division of roles between the actors, or by creating a new division of roles. The idea behind this is that if one of the means of influence disappears (e.g. the municipality no longer has the possession of the land as a resource) this can be compensated by using other means of influence (e.g. rules or ideas). This means that we are looking for tools which enable the power balances in the development process to be changed. The distinction between active and passive land policy allows the notion of tools to mean two things. It refers to 'instruments' by which the planning authority can influence the behaviour of others that have taken the initiative for development, thus acting indirectly upon the course of the development process. It also refers to ways of proceeding in which the planning authority takes the initiative and plays a direct role in the

256

development process, for example by actively buying the land before development is to take place as we mentioned before. Table 6.1 shows the options for using such tools. Here we reconsider the findings from the case studies once more, to take a step further from this table and describe what a planning authority could do legally to pursue each of the three approaches to influence expenditure on the residential environment.

In the first of the three approaches mentioned above, i.e. the creation of legal powers to make clear in advance what private actors have to contribute, the planning authority uses its public powers. Tools under public law are used, by which the planning authority imposes upon the private actors what they should do. Examples shown in the table are expropriation and pre-emption rights to obtain the land from the first landowners without leaving them with a financial margin, or the tools of *Baulandumlegung* and *Erschliessungsbeiträge* which impose contributions upon either the first or the final landowers. As can be seen in the table, these are tools that fit into a hierarchical approach to housing development.

The approach of avoiding the creation of legal rights before agreement with the developers are reached about their contribution to the development, implies in any case that private and public partners interact before legal rights concerning the development exist. This does not comply with the use of tools under public law to regulate the interactions, because using such tools would involve creating legal rights between the partners. Before such rights exist, the relations between the planning authority and the private partners are regulated under private law. Referring to table 6.1, amicable land acquisition and public private-partnership are tools that can be used in this case.

This is not the whole story, however. These last tools work in an environment where the market is the main coordinating principle. But part of the solution of avoiding the creation of legal rights before negotiations with the private partners are realised is to reach an agreement. This involves a network as coordinating principle, in which relations between the parties involved are arranged in contracts. In that respect, the second and the third approach described above have similar aspects. The difference is that if no legal rights have been created before, the planning authority can use the possibility of creating them as 'loose change' in the negotiations with the private actors. We will call this approach 'induced cooperation'. In section 6.4 we return to how this could be used to change the development process. This comes close to the approach to influence the residential environment described above, i.e. working closely together with the private actors from the start of the development process. But in this last approach, the legal rights are not necessarily used to reach 'induced cooperation'. Cooperation can also be based on a mutual understanding of the advantages of working together in the realisation of the housing development. This can be realised various forms of public private partnership.

To translate the descriptive table 6.1 into prescriptive statements, we need to apply it to a concrete situation because such prescriptions are always context-dependent. Therefore, in the next section, we take the case of the Dutch land market. We apply the argument developed here to this particular case, to show how it can be used to discern and explore possible policy options.

6.4 Policy options for the Dutch situation

This study started with the observation that in the Netherlands, the situation concerning land policy and housing development has changed considerably since the beginning of the 1990s. What can be said on the basis of the analysis reported in this book for the Dutch situation? This is at the same time an application of the general results of this study to a concrete problem. As such, it is not only aimed at Dutch readers. It can be useful to all readers interested in how the aproach developed here can be used to influence the practice of land policy and housing development. It is also an illustration of how a country can benefit from experiences in other countries.

Land policy and housing development in the Netherlands

To understand the current Dutch problems concerning housing development and land policy we have to go back in history, to the beginning of the 1950s. In this period just after the second world war, the need for houses was urgent in the Netherlands, due to damage to the existing housing stock and a delay in the housing production during the years of the war plus a high population growth. For the municipalities, this was a reason to get actively involved in the process of land development. From this moment onwards, the Dutch planning policy generally aimed at designating sufficient building land in official plans to ensure that there was no shortage of land for housing (or industry, or offices). Land policy aimed at ensuring that the land designated in official plans was offered for development in the form of serviced building plots. This was the policy and practice for many decades. In addition, it was public policy that housing should be built to satisfy the need, even when those needing housing could not afford to pay the commercial cost of it. The result was that government stimulated the building of housing for those who could afford it, and subsidised or otherwise supported the building of housing for those who needed it but could not afford it. Land and planning policy ensured that sufficient building land was available for house building. The result was that (marketable) houses sold for prices that were only a little above their production costs (see also Needham, 1997).

It is, therefore, no surprise that private developers did not want to be involved in the land development process, it was not profitable. A division of labour had grown up, whereby the municipality undertook the land development and accepted the risks during the five to ten years between land acquisition and land disposal. Private developers constructed buildings on the serviced plots and accepted the risks during a much shorter period. Moreover, the planning process gave a high degree of certainty to the building developer. The result was that for developers, construction was a fairly predictable production process on which they could make normal profits (comparable to making books, or furniture, or cars): They did not want, nor need to get involved in land development. The results of the Dutch land policy 'on the ground' during the last decades have often been qualified as satisfactory (see for example Badcock, 1994). All the parties seemed happy with the situation.

However, it did not last. Since the beginning of the 1990s, important changes in the market for housing land have taken place, which affect the housing development process. House prices have risen and the share of social housing in all new housing developments has fallen. As a result, the profits to be made from developing a housing estate have increased greatly.

Commercial developers have not waited for municipalities, but have themselves bought unserviced land which is intended for development. The suppliers of that land, well advised regarding its value, have often sold it for many times its existing value.

The municipalities have been deprived of their usual way of recouping the costs of public facilities (by selling serviced land with development rights). And they have discovered that the legislation does not allow them to recoup those costs adequately in other ways. In particular, the local land use plan (the *bestemmingsplan*) is legally binding, which means that the owner of the land has the right to develop it in accordance with the plan. This right cannot be withheld by the municipality and made subject to a negotiated agreement about who pays what costs. In the space of a few years, municipalities have moved from a situation in which they had an extremely powerful instrument to arrange the expenditure on the residential environment, to a situation in which they only have a few, weak instruments for doing that. Moreover, municipalities have not been prepared for this change: Legislation has not been adapted and the municipalities have not had any experience in negotiating about expenditure on the residential environment.

Lessons to be learnt

Placed within the framework of table 6.1, the recent changes in the Netherlands can be described as a forced shift from an active to a passive land policy. This situation is considered by municipalities as problematic because the land policy measures available are not adapted to this situation. Basically, there are two ways out of this situation, either by adapting the land policy measures to this new situation of a passive role of the municipalities on the land market, or by creating possibilities for the municipalities to return to the 'old' practice of active involvement on the land market. Using again table 6.1, there are three options in pursuing either of these possibilities. Below, we deal briefly with each of these options.

If the choice is made to accept a passive role of the municipalities on the land market, then the following options are open:

- Market as coordinating principle: The only option for using the market as a coordinating principle without being directly active on that market is by entering into public-private partnerships. The repartition of tasks and responsibilities between the public and private partners will be subject to negotiations and can be fixed in a contract. Thus, the municipalities could exercise a close supervision over the activities on the market, without being directly active on that market (see the case of Rennes la Poterie).
- Network as coordinating principle: To describe the way in which networks could be used as a coordinating principle we have introduced the notion of 'induced cooperation' in section 6.3. By this we mean procedures that "induce" private developers to enter into negotiations with the municipalities for the realisation of their development projects. In the cases of Cramlington North-East Sector and Bishop's Cleeve, this option was used by means of planning gain agreements. The municipalities withheld their planning permission until an agreement was reached with the private developers about their contribution to the residential environment. Thus, negotiations between the planning authorities and the developers were induced. In Stuttgart Hausen-Fasanengarten, the idea of induced cooperation was used in a very creative way. The municipality would not change the land use plan unless the landowners agreed to take care of the land development by themselves, and to make several other contributions to the residential environment. Thus, the first landowners were induced into cooperation with the municipality.
- Hierarchy as coordinating principle: This means introducing tools that allow expenditure on the residential environment to be recouped from the developers or the first landowners. Examples of such tools can be found in other countries. The German *Erschließungsbeiträge* are an example (see the case of Bonn Ippendorf). In the Netherlands, central government is currently working on the development of such a tool, the *exploitatieheffing*.

If the passive role of the municipalities on the land market is not accepted, and they try to return to an active land policy, then this has to be done in one of the following ways:

261

– Market as coordinating principle: Municipalities could start to buy unserviced building land again. Given the higher land prices, the municipalities would need more money for this. At the same time, policy measures could be envisaged to reduce the price of acquisition (by municipalities) to be paid to the first landowners. To this aim, it has to be clear to the buyers of unserviced building land what financial contributions they have to make to the residential environment. They will then calculate this – through a residual calculation, see chapter three – into the price they are prepared to pay for the land. Another option would be to return to the old situation of an abundant supply of building land. However, this does not seem realistic because of present political concerns regarding the preservation of open spaces.

– Network as coordinating principle: Here again, the notion of induced cooperation applies. A condition for this to work is that the private developers depend on the municipalities for the realisation of their plans. This dependence usually exists because a development cannot be started unless the municipality grants a planning or a building permit. This can be used by the municipality to 'induce' cooperation. The French procedure of the *Zone d'Aménagement Concerté* (ZAC), as it was used in Bois-Guillaume Portes de la Forêt and in Rennes la Poterie is an example of such induced cooperation. The phases of creation and of realisation of the development plan are separated in this procedure. The plan enters into the phase of realisation only when the parties involved agree upon the conditions under which it is to take place.

– Hierarchy as coordinating principle: Using hierarchy as coordinating principle to enable municipalities to pursue an active land policy means enforcing the tools of expropriation and pre-emption. Thus, the power balance between municipalities and private actors could be changed. In the terms of our analysis, the lack of resources of the municipalities which results in them not being able to buy unserviced building land would then be compensated by more rules. A discussion about how pre-emption and expropriation rules might be changed to make them more effective is taking place in the Netherlands.

All of these options have their strengths and their weaknesses in particular situations. The analysis reported in this book can give ideas about which option to use under which circumstances. Because of the importance of the context of each development process it is precarious to make any general recommendations. The choice of how to react to the changed circumstances on the Dutch land market is a political one. This choice should always be based on a (normative) perception of the role of the (local) government and on an appreciation of the context of each development process. When these two items are clear, the choice for the most appropriate policy option, as presented above, follows more or less naturally. However, whether or not it can be put into practice depends again on political preference, because each of the policy options requires money, or changes in legal rules, or both.

6.5 Where to go from here?

To complete this book, a brief reflection upon the broader significance of the results of this study is appropriate. This goes beyond what can be concluded strictly on the basis of the case studies carried out here. The process of housing development is used here as an example of a policy process in which various actors play a role, and in which private and common interests have to be reconciled. As such, the eight development processes of greenfield housing studied here can be seen as a laboratory in which different mixtures of economic and institutional aspects; of resources, rules, and ideas; of markets, hierarchies, and networks; or in a more general sense, of state and market in policy processes have been investigated. The primary concern of this investigation was to generate insight into the course and the outcomes of housing development processes on greenfield sites. But it gave some clues as to how in a more general sense, public and private actors can work together effectively in policy processes. We end this book with some observations on this subject, which pose as many new questions as they provide conclusions.

With hindsight, an important observation that can be made on the basis of this research is that within mixtures of public and private actors, equilibirium conditions seem to exist in which the power balances appear to be satisfactory – or in any case acceptable – to each of the actors involved. In several cases, e.g. in Stuttgart Hausen-Fasanengarten, in Bois-Guillaume Portes de la Forêt, in Rennes la Poterie, and in Zwolle Oldenelerbroek, each of the actors encountered claimed to be satisfied, or even happy with the way things had gone. The observation of equilibrium conditions in this study does not stand on its own. In an investigation of the relationship between the development industry and the planning system in the United Kingdom, Healey (1998a) describes how this relationship develops and changes in time. What occurs is that after this relationship has gone through a sudden change, due to changes in context variables, such as the market conditions, or the political ideology, it tends to find a new equilibium in which the different parties can work together satisfactorily.

The description in section 6.4 of the recent changes in the Dutch situation can be understood in this light. Because of a sudden shift towards a more market led development process, the land development in the Netherlands has suddenly changed from a public activity into a private activity, thus showing how equilibrium states can be vulnerable. The institutional structure, that is, the relationship between the public (municipality) and private (house builders) actors had been built up over the years on a completely different basis. Therefore since the mid 1990s, a revision of the roles and instruments of state and market in the housing development process is in process. What instruments or working methods do the municipalities need to successfully accompany private land development? This can be described as the search for a new equilibrium.

In France, at the beginning of the 1990s, a situation occurred that can be seen in some respects as the opposite to the Dutch. As a result of a crisis in the market for real estate – especially in the Paris region – market-led development processes came to a standstill. The lack of demand for real estate induced the private developers to postpone the realisation of their development projects. At the same time, as a result many public facilities, which were part of the development schemes, were not realised. This provoked a discussion about whether private developers should be responsible for the land development, and how public parties could break the deadlock if the private developers did not want to develop. While in the Netherlands it was an increasing demand for building land and real estate compared to the supply, in France it was a sudden fall in the demand for real estate compared to the supply which provoked the changes. In both cases, the result can be described as the overthrowing of an equilibrium between the parties involved in the development process of real estate, i.e. the development industry and the planning authorities.

Combining our experience of this study with the investigation by Healey, we are inclined to draw the conclusion that equilibrium states in the development process are generally beneficial to all of the parties involved. To support this assertion, we use the notion of 'institutional capacity building' (see Healey, 1997, 1998a, 1998b). In Healey's words, the concept of institutional capacity '...refers to the overall quality of the collection of relational networks in a place. (...) There is an increasing recognition that the quality of this capacity matters, whether the collective objective is economic competitiveness, sustainable development, biospheric sustainability or quality of life' (1997: 61). Although according to Healey the importance of institutional capacity is recognised, its contents remain rather indistinct. '...a key element of such capacity lies in the quality of local policy cultures. Some are well integrated, well connected, and well informed, and can mobilise readily to capture opportunities and enhance local conditions. Others are fragmented, lack connections to sources of power and knowledge, and the mobilisation of capacity, to organise to make a difference' (Healey, 1998b: 1531).

Such criteria for institutional capacity are difficult to make explicit. This is a problem if we want to turn our attention to the building of institutional capacity. To do so it has to be clear how it can be recognised. If we know what makes the institutional capacity in one case higher than in the other, we would have a clue to enhance such capital in other cases. We want to propound the idea that power balances between the actors involved in a development process can be used as a tool to analyse and predict the institutional capacity in that process. It is our hypothesis that institutional capacity building can be understood in terms of the search for an

265

equilibrium in the power balances between the parties involved in development processes. This hypothesis was not part of the analytical framework that has been applied in the cases. Therefore we can support it only by a theoretical argumentation which is given below.

Power, ideas and 'enrichment' in policy processes

As argued in section 4.4, power is not considered in this study as something one actor has to the extent that the other actor over whom it is wielded does not have it. Power can be generated in a development process. Theoretically, this opens the possibility for a continuous increase in institutional capacity, allowing the outcomes of the development process to be 'enriched' (see Teisman, 1992). Before explaining this, it has to be clear what is meant by the notion of 'enrichment'. Put simply, this is the extra value that can be obtained when various actors in a policy process are able to add their ideas to an initial policy proposal. In that case, the process and its outcomes will be judged by all the participants to be of higher quality than the initial proposal, because the process and its outcomes will have been approached and constructed from a number of perspectives. The generation of power in development processes can lead to such enrichment because when an actor has more power, he has better possibilities to reach his objectives, or to realise his ideas. If all of the actors have that, more objectives are likely to be reached, and more ideas are likely to be realised. This would mean an 'enrichment' of the outcomes compared to what could have been reached in the first instance.

Such 'enrichment' is in our view the objective of institutional capacity building. What else could be the use of building institutional capacity if it was not to lead to a process and outcomes which are judged to be of higher quality than what would otherwise have been realised? This can only be reached if the policy process is open in the sense that all the ideas brought in to the process are considered seriously. This can only be the case if there is not one – or a limited number – of actors that 'have the power'. If one or a few actors have an influence on the process and its outcomes, leaving no room for the ideas and objectives of others to be taken into consideration, 'enrichment' will not take place. Both Teisman and Healey want to reach this situation by promoting collaborative (Healey) or communicative

(Teisman) planning processes. The idea behind this is that if all the actors communicate their wishes and concerns, and collaborate to reach them, the outcome of the process will be optimal in terms of the satisfaction of the actors involved. And since in the ideal collaborative and communicative planning process, all stakeholders are involved, this is the optimal outcome of the process. Our 'balanced interdependence' view of institutional capacity as a condition for enrichment can be used to qualify this.

The dependence or power relations can be asymmetrical, and nevertheless in balance. This is the case when the power of the actors enables them to reach their objectives. This was the situation in the Netherlands before the changes on the land market in the 1990s. Although the power of the municipalities was much bigger than that of the other parties in the development process, their position was not contested because each actor could reach his objectives. There was an equilibrium state and in our view at that time the institutional capacity was high (as in the cases of Zwolle Oldenelerbroek and of Arnhem Rijkerswoerd). It is, therefore, not necessary in order for development processes to be enriched, that all stakeholders have equal access to and influence on the development process, as sometimes seems to be the assumption when there is talk of communicative or collaborative planning processes. It is enough if there is an equilibrium in the power relations in the sense described above. This has not been analysed in these terms in this study. But although it is only indirectly based on this study, we think it is a line along which the ideas developed here could be carried further with the aim of gathering insight in the relationship between the development process and its outcomes.

We can make yet another assertion about a possible option to enable development processes to be 'enriched'. To that aim, we refer to the quote from Forester we used in the first chapter: 'What if social interaction were understood neither as a resource exchange (micro-economics) nor as an incessant strategising (the war of all against all), but rather as a practical matter of making sense together in a politically complex world?' (1993: ix). In this study, the resource exchange (the financial analysis) and the strategising (the actors and how they aim at realising their objectives) have been analysed extensively. Although our focus was on how in interactions the interdependent actors decide about (expenditure on) the residential environment, the two 'driving forces' of financial considerations and individual strategising remain obvious. In our attempt to combine the two and to describe the process as 'a practical matter of making sense together in a politically complex world' one element of our analysis attracts attention. It is the element of the ideas, and the role they play in housing development. We have seen in the cases of Bois-Guillaume Portes de la Forêt and Stuttgart Hausen-Fasanengarten that ideas allowed unexpected outcomes to be reached and more than that, in these cases the ideas made all the actors feel part of an achievement. This binding force of ideas might be worth exploring in the search for ways to improve the functioning of development processes.

References

Abma, T.A. (1996), *Responsief evalueren: Discoursen, controversen en allianties in het postmoderne*, Delft: Eburon.

Acosta, R., V. Renard (1994), *European urban land and property markets in France*, European urban land and property markets 3, London: UCL Press.

Albrechts, L. (1991), Changing roles and positions of planners, *Urban Studies*, vol. 28, no.1, 1991, pp. 123-137.

Allison, G.T. (1972), *The essence of decision: Explaining the Cuban missile crisis*, Boston: Little Brown.

Badcock, B. (1994), The strategic implications for the Randstad of the Dutch property system, *Urban Studies*, vol. 31, pp. 425-445.

Balchin, P. (ed.) (1996), *Housing policy in Europe*, London and New York: Routledge.

Ball, M. (1998), Institutions in British property research: a review, *Urban Studies*, vol. 35, no. 9, pp. 1502-1517.

Barlow, J., S. Duncan (1994), *Success and failure in housing provision: European systems compared*, Oxford: Pergamon.

Barret, S., P. Healey (1985), Land policy: Problems and alternatives, Aldershot: Gower.

Blowers, A. (1980), The limits of power, Oxford: Pergamon Press.

Bramley, G. (1993), Land-use planning and the housing market in Britain: the impact on housebuilding and house prices, Environment and Planning A, no. 25, pp. 1021-1051.

Burie, J.B. (1972), Wonen en woongedrag: verkenningen in de sociologie van bouwen *en wonen*, Meppel: Boom.

Carmona, M. (1999), Controlling the design of private sector residential development: an agenda for improving practice, Environment and Planning B: Planning and Design, vol. 26, pp. 807-833.

Cassel, P. (ed.) (1993), The Giddens reader, Stanford: Stanford University Press.

Chambert, A. (1988), The dynamics of urban development: Building, housing, and planning in the Stockholm region 1950 - 1980, paper presented at the seminar 'Urban politics and urban development', University of Utrecht, 25 October.

Coase, R.H. (1937), The nature of the firm, Economica, vol. 4, pp. 386-405.

Comby, J., V. Renard (1996), Les Politiques Foncières, Que sais je, vol. 3143, Paris: Presses Universitaires de France.

Crozier, M., E. Friedberg (1977), L'acteur et le système, Paris: Editions du Seuil.

David, C.H. (1995), German urban planning and building laws – Deutsche Städtebauvorschriften, Dortmunder Beiträge zur Raumplanung 71, Dortmund: Institut für Raumplanung (IRPUD).

Davy, B. (1999), Land values and planning law: the German practice, Presentation held at the XIII AESOP Congress in Bergen, Norway.

Dekker, A., H. Goverde, T. Markowski, M. Ptaszynska-Wolockzkowicz (1992), Conflict in urban development, Aldershot: Ashgate.

Dieterich, H. (1996), Baulandumlegung, 3, Auflage, München: Verlag C.H. Beck.

Dieterich, H., E. Dransfeld, W.Voss (1993a), Urban land and property markets in Germany, European urban land and property markets 2, London: UCL Press.

Dieterich, H., E. Dransfeld, W. Voss (1993b), Funktionsweise städtischer Bodenmärkte in Mitgliederstaaten der Europäischen Gemeinschaft: ein Systemvergleich, Bundesministerium für Raumordnung, Bauwesen und Städtebau, Pulheim: Lottmann.

Dieterich, H., E. Dransfeld (not dated), Hohe Baulandpreise: Naturgesetz oder hausgemacht? FORUM – Magazin für junges bauen.

Douma, S., H. Schreuder (1991), Economic approaches to organisations, Hemel Hempstead: Prentice Hall.

Edwards, M. (1995), Agents and functions in urban development, Cartas Urbanas, no. 4, pp. 26-39.

Elias, N. (1971), Wat is sociologie ?, Utrecht: Het Spectrum.

Ennis, F. (1997), Infrastructure provision, the negotiating process and the planner's role, Urban Studies, vol. 34, no. 12, pp. 1935-1954.

European Commission (1999), The EU compendium of spatial planning systems and policies: Germany, Luxembourg: Office for official publications of the European Communities.

Evans, A.W. (1983), The determination of the price of land, Urban Studies, vol. 20, pp. 119-129.

Evans, A. W. (1987), House prices and land prices in the South-East: a review, London: The House builders Federation.

Evans, A.W. (1991), 'Rabbit hutches on postage stamps': Planning, development and political economy, Urban Studies, vol. 28, no. 6, pp. 853-870.

Eve, G. (1992), The relationship between house prices and land supply, Department of the Environment, Planning research programme, London: HMSO.

Fischer, J. (1992), Integrating research on markets for space and capital, Journal of AREUED, 20, pp. 161-180.

Fischer, F., J. Forester (eds.) (1993), The argumentative turn in policy analysis and planning, Durham and London: Duke University Press.

Flyvbjerg, B. (1998), Rationality and power: Democracy in practice, Chicago: The University of Chicago Press.

Forester, J. (1993), Critical theory, public policy and planning practice: Toward a critical pragmatism, Albany: State University of New York Press.

Galtung, J. (1977), Methodology and Ideology: Essays in methodology volume one, Copenhagen: Christian Ejlers.

Galtung, J. (1990), Theory formation in social research: a plea for pluralism, in: Øyen, E. (ed.), Comparative methodology: Theory and practice in international social research, London: Sage Publications Ltd.

George, H. (1879), Progress and poverty: an inquiry into the cause of industrial decline and of increase of want with increase of wealth: the remedy, reprint 1890, London: Kegan Paul.

Giddens, A. (1984), The constitution of society: Outline of the theory of structuration, Cambridge: Polity Press.

Glaser, B.G., A.L. Strauss (1967), The discovery of the grounded theory, Chicago: Aldine.

Goldsmith, M. (1993), Local Government, International Perspectives in Urban Studies 1, edited by R. Paddison e.a., University of Glasgow.

Gore, T., D. Nicholson (1991), Models of the land development process: a critical review, Environment and Planning A, vol. 23, pp. 75-730.

Gravier, J.F. (1947), Paris et le désert français, Paris: Flammarion.

Greef, J.H. de (1997), Het gevecht om het residu, Amsterdam: Universiteit van Amsterdam.

Guba, E.G. (1990), The alternative paradigm dialog, in: Guba, E.G. (ed.), The paradigm dialog, London: Sage Publications.

Hajer, M.A. (1995), The politics of environmental discourse: Ecological modernisation and the policy process, Oxford: Clarendon Press.

Hall, P.A., R.C.R. Taylor (1996), Political science and the three new institutionalisms, Political Studies, vol. XLIV, no. 5, pp. 936-957.

Healey, P. (1992a), The reorganisation of state and market in planning, Urban Studies, vol. 29, nos. 3/4, pp. 411-434.

Healey, P. (1992b), An institutional model of the development process, Journal of Property Research, no. 9, pp. 33-44.

Healey, P. (1997), Collaborative planning: Shaping places in fragmented societies, London: Macmillan Press Ltd.

Healey, P. (1998a), Regulating property development and the capacity of the development industry, Journal of property research, vol. 15, no. 3, pp. 211-227.

Healey, P. (1998b), Building institutional capacity through collaborative approaches to urban planning, Environment and Planning A, vol. 30, pp. 1531-1546.

Healey, P., S. M. Barret (1990), Structure and agency in land and property development processes: Some ideas for research, Urban Studies, vol. 27, no. 1, pp. 89-104.

Healey, P., M. Purdue, F. Ennis (1995), Negotiating Development: rationales and practice for development obligations and planning gain, E & FN SPON, London.

Huberts, L., M. de Vries (1995), Case studies en besluitvormings-onderzoek: mythen en mogelijkheden, in: P. t'Hart, M. Metselaar, B. Verbeek, Publieke Besluitvorming, 's-Gravenhage: Vuga Uitgeverij.

Janssen, J., B. Kruyt, B. Needham (1994), The honeycomb cycle in real estate, The journal of real estate research, 9, pp. 237-251.

Jegouzou, Y. (ed.) (1997), Urbanisme, Dalloz Action, Paris: Dalloz.

Kam, G. De (1996), Op grond van beleid: locaties voor de sociale woningbouw, grondbeleid en ruimtelijke spreiding van welstand in en rond Den Haag, Almere: Nationale Woningraad.

Korsten, A.F.A., A.F.M. Bertrand, P. de Jong, J.M.L.M. Soeters (1995), Internationaal vergelijkend onderzoek, 's-Gravenhage: VUGA.

Korthals Altes, W.K. (1998), Grondbeleid als marktinterventie, Intreerede, Faculteit Civiele Techniek en Geowetenschappen, Delft: Technische Universiteit Delft.

Krabben, E. van der (1995), Urban Dynamics: a real estate perspective, an institutional analysis of the production of the built environment, PhD Thesis, Tilburg University, Center for Economic Research.

Kruyt, B., Needham, B., Spit, T. (1990), Economische grondslagen van grondbeleid, Stichting voor beleggings- en vastgoedkunde.

Kuiper Compagnons (1990), Kwaliteitskosten signalering grond, Rotterdam/Arnhem.

Kuiper Compagnons (1991), Basiseisen woonomgeving; methodiek uitwerking basiseisen, Rotterdam/Arnhem.

Lacaze, J.P. (1995), Introduction à la Planification Urbaine: Imprécis de l'urbanisme à la française, Paris: Presses de l'école nationale des ponts et chaussées.

Lambooy, J.G. (1990), Urban land and land prices: an institutional approach, The social nature of space, edited by B. Hamm & B. Jalowiecki, Polish Academy of Sciences, Committee for Space Economy and Regional Planning, Warzawa.

Lipsey, R.G. (1966), An introduction to positive economics, 2nd edition, London: Weidenfeld and Nicholson.

Magalhaes, C. de (1999), Social agents, the provision of buildings and property booms: The case of São Paulo, Journal of urban and regional research, vol. 23 (3), pp. 445-463.

Martinand, C., J. Landrieu (1996), L'aménagement en questions, Paris: adef.

Mastop, H., A. Faludi (1997), Evaluation of strategic plans: the performance principle, Environment and Planning B: Planning and Design, vol. 24, pp. 815-832.

Mill, J.S. (1849), Principles of political economy with some of their applications to social philosophy, 2nd. Edition, London.

Mills, E.S. (1972), Urban economics, Glenview: Scott, Foresman and Company.

Ministerie van VROM (1996), Kwaliteitswijzer woonomgeving deel A: Kavel en blokniveau, deel B: Buurt en wijkniveau, Zoetermeer: Distributiecentrum VROM.

Mitchel, C. (1983), Case and situation analysis, Sociological Review, vol. 31, no. 2, pp. 187-211.

Monk S., B. Pearce, C. Whitehead (1991), Planning, land supply and house prices: a literature review, Land economy monograph 21, Department of land economy, University of Cambridge.

Muller, P., Y. Seurel (1998), L'analyse des politiques publiques, Paris: Montchrestien.

Muth, R.F. (1960), The demand for non-farm housing, The demand for durable goods, in: Harberger, A.C. (ed.), Chicago III, The University of Chicago Press.

Needham, B. (1997), Land policy in the Netherlands, TESG, vol. 88, pp. 291-296.

Needham, D.B. (1992), A theory of land prices when land is supplied publicly: the case of the Netherlands, Urban Studies, vol.29, no.5, pp. 669-686.

Needham, D.B. (1998), Housing and land in Israel and the Netherlands: a comparison of policies and their consequences for the access to housing, Town Planning Review, vol. 69, no. 4, pp. 397-423.

Needham, D.B., P. Koenders, B. Kruijt (1993), Urban land and property markets in the Netherlands, European Urban land and property markets 1, London: UCL Press.

Needham, D.B., R.W. Verhage (1998), The effects of land policy: quantity as well as quality is important, Urban Studies, vol. 35, no. 1, pp. 25-44.

Neutze, M. (1987), The supply of land for a particular use, Urban Studies, vol. 24, pp. 379-388.

Newman, P., A. Thornley (1996), Urban planning in Europe: International competition, national systems and planning projects, London and New York: Routledge.

Nooteboom, L.W. (1994), Nederlandse grondverwerving en gronduitgifte: een huwelijk onder druk, Nijmegen: Katholieke Universiteit Nijmegen.

Ostrom, E. (1986), A method of institutional analysis, in: F.X. Kaufman, G. Majone, V. Ostrom (eds.), Guidance, control and evaluation in the public sector, New York, Berlin: Walter de Gruyter.

Øyen, E. (1990), Comparative Methodology: Theory and practice in International Social Research, London: Sage Publications.

Priemus, H. (1998), Housing research and the dynamics of housing markets, in: A.J.H. Smets, T. Traerup (eds.), Housing in Europe: Analysing patchworks, Utrecht/Hørsholm: Utrecht University/Danish Building Research Institute.

Renard, V. (1988) Financing public facilities in France, in: R. Alterman (ed.), private supply of public services, pp. 173-181, New York: New York University Press.

Renard, V. (1996), Quelques caractéristiques des marchés fonciers et immobiliers, in: Economie et Statistique, no. 294-295, pp. 89-97.

Ricardo, D. (1812), The principles of political economy and taxation, reprint with introduction by David Winch, 1937, London: J.M. Dent & Sons.

Rosen, K. (1983), Towards a model of the office building sector, Journal of Areued, 15(4), pp. 281-299.

Teisman, G.T. (1992), Complexe besluitvorming: een pluricentrisch perspektief op besluitvorming over ruimtelijke investeringen, 's-Gravenhage: VUGA.

Teisman, G.T. (1997), Sturen via creatieve concurrentie: Een innovatie planologisch perspectief op ruimtelijke investeringsprojecten, Nijmegen: Katholieke Universiteit Nijmegen.

Thompson, G., J. Frances, R. Levacic, J. Mitchell (eds.) (1991), Markets, hierarchies and networks: the coordination of social life, London: Sage Publications.

Verhage, R. (1998), Wie betaalt de woonomgeving na de Vinex?, Stedebouw en Ruimtelijke Ordening, 1998/4, pp. 19-48.

Verhage, R., B. Needham (1997), Negotiating about the residential environment: It is not only money that matters, Urban Studies, vol. 34, no. 12, pp. 2053-2068.

Wester, F. (1995), Strategieën voor kwalitatief onderzoek, 2nd edition, Bussum: Coutinho.

Williams, R.H., B. Wood (1993), Urban land and property markets in the United Kingdom, European urban land and property markets 4, London: UCL Press.

Willig, R.D. (1987), Contestable markets, in: J. Eatwell, M. Millgate, P. Newman (eds.), The new Palgrave: A dictionary of economics, pp. 618-622, London: MacMillan.

Winter, J., T. Coombes, S. Farthing (1993), Satisfaction with space around the home on large private sector estates, Town planning review, 64 (1), 1993, pp. 65-87.

Zwanikken, T. (2001), Ruimte als voorraad? Ruimte voor variëteit!, PhD Thesis Katholieke Universiteit Nijmegen: Faculteit der Beleidswetenschappen.

Documentation of the case studies

Arnhem Rijkerswoerd phase two

Bouwfonds Nederlandse Gemeenten N.V. (1988), *Overeenkomst inzake de bebouwing van een gedeelte van de tweede fase van het plan Rijkerswoerd*, behorend bij raadsvoorstel I-156, Gemeenteraad Arnhem, 16 mei 1989.

Gemeente Arnhem, *Vragenlijst selectie tweede fase Rijkerswoerd.*

Gemeente Arnhem (1985), *Bestemmingsplan Rijkerswoerd.*

Gemeente Arnhem (1986), *Overwegingen bij de aanpak tweede fase Rijkerswoerd*, internal note.

Gemeente Arnhem (1987), *Theoretische uitgangspunten Rijkerswoerd fase twee.*

Gemeente Arnhem (1988), *Rijkerswoerd tweede fase: randvoorwaarden en richtlijnen.*

Gemeente Arnhem (1992), *Rijkerswoerd: Wonen tussen groene vingers.*

Gemeente Arnhem (1993), *Herziening grondexploitatie Rijkerswoerd: wonen.*

Zwolle Oldenelerbroek

Gemeente Zwolle (1988), *Exploitatie-berekening Zwolle Zuid*, maart 1988.

Gemeente Zwolle (1991), *Bestemmingsplan Oldenelerbroek.*

Gemeente Zwolle (1992), *Exploitatie-opzet Zwolle Zuid*, mei 1992.

Gemeente Zwolle (1996), *Overzicht grondverkopen Zwolle Zuid*, internal note.

Stichting Woningbeheer Zwolle (1995), *Info nieuwbouwwoningen.*

Stichting Woningbeheer Zwolle (1996), *Wonen in Oldenelerbroek*, Zwolle Zuid, Info-bulletin.

Bishop's Cleeve

Agreement made the third day of August 1987, between: Tewkesbury Borough Council, Severn-Trent Water Authority, Bovis Homes Ltd., Robert Hitchins Builders Ltd. ('the developers'), and 'the owners'.

Agreement made the third day of August 1987, between: 'the owners', 'the developers', and Gloucestershire County Council.

Agreement made the twenty-first day of May 1991, between Tewkesbury Borough Council, Bovis Homes Ltd., Robert Hitchins Ltd., Gloucester Land Company Ltd.

Gloucestershire County Council (1980), *The structure plan for Gloucestershire*.

Tewkesbury Borough Council (1984), *Cheltenham environs local plan: Report of a public inquiry into objections to the plan*.

Tewkesbury Borough Council (1985), *Outline permission for development*, Residential (approx. 1,000 dwellings) development on land adjoining and to the west of Bishop's Cleeve.

Tewkesbury Borough Council (1986), *Cheltenham environs local plan: written statement*.

Tewkesbury Borough Council (1994), *Tewkesbury Borough local plan*, draft for consultation.

Cramlington North-East Sector

Agreement made the twenty-ninth day of March 1974, between: Northumberland County Council, Seaton Valley District Council, William Leech (Investment) Ltd., Cramlington Developments Ltd., William Leech (Builders) Ltd., John T. Bell & Sons Ltd.

Bellway Homes (North-East) Ltd. (1994), *Town and country planning act 1990: section 78 public inquiry into proposed residential developments at North-East sector of Cramlington 1d*, Appendices to the proof of evidence of Nigel Perry.

278

Blyth Valley District Council (1982), *Cramlington North-East residential sector: Housing strategy*, Planning and development services committee.

Blyth Valley District Council (1991), *Key Statistics*.

Blyth Valley District Council (1994), *Blyth Valley District local plan*, draft for consultation.

Northumberland County Council (1965), *Cramlington comprehensive development area: written analysis*, County development plan.

Northumberland County Council (1970), *Cramlington comprehensive development area: written statement*, County development plan.

Northumberland County Council (1979), *Northumberland County structure plan: written statement*.

Taylor, H.A. (1963), Cramlington: *Historical survey, the Northumberland County archivist*.

Bonn Ippendorf

Stadt Bonn (1985), *Bestandsverzeichnis/Bestandskarte Umlegungsverfahren 211*.

Stadt Bonn (1985), *Bebauungsplan nr. 7618-14 der Stadt Bonn*.

Stadt Bonn (1989), *Bebauungsplan nr. 7618-15 der Stadt Bonn*.

Stadt Bonn (1990), *Umlegungsplan 211*.

Stadt Bonn (1990), *Zahlenangaben und Daten zum Umlegungsverfahren 211*, internal note.

Stuttgart Hausen-Fasanengarten

Erschließungsvertrag Erweiterung Hausen 1 und Hausen 2 (1995), zwischen: Landeshauptstadt Stuttgart und Erschließungsgemeinschaft Erweiterung Hausen 1 und Hausen 2.

Gesellschaft für Stadt und Landesplanung MBH (1996), *Gesellschaftsvertrag Erschließungsgemeinschaft Fasanengarten 1 und 2*, vertreten zwischen Eigentümern.

Gesellschaft für Stadt und Landesplanung MBH (1996), *Bauland-umlegung als Handlungsfeld kommunalen Flächenmanagements*, Vortrag im 350. Kurs des Instituts für Städtebau Berlin.

Landeshauptstadt Stuttgart (1993), *Durchführung von freiwilligen Umlegungen im Gebiet Weilimdorf - Erweiterung Hausen 1 und 2*, Gemeinderatsdrucksache 313/1993.

Landeshauptstadt Stuttgart (1993), *Bebauungsplan Erweiterung Hausen 1*.

Landeshauptstadt Stuttgart (1995), *Auswirkung der besonderen Bedingungen bei freiwilligen Umlegungen*, Anlage für die Mitglieder des Ausschusses für Bodenordnung.

Landeshauptstadt Stuttgart (1996), *Bebauungsplan Erweiterung Hausen 2: Satzungsbeschluß*, Gemeinderatsdrucksache nr. 133/1996.

Landeshauptstadt Stuttgart (1998), *Wirtschaftlichkeitsbetrachtung der Umlegungen Hausen-Fasanengarten 1 und 2*, internal note, Amt für Stadterneuerung.

Landeshauptstadt Stuttgart (1998), *Ausdruck aus der Umlegungs-Statistik-Datei*, Hausen-Fasanengarten 1 und 2.

Vertrag Freiwillige Umlegung im Gebiet Weilimdorf - Erweiterung Hausen 1 (1993), zwischen Landeshauptstadt Stuttgart und Gesellschaft für Stadt und Landesplanung mbH.

Städtebaulicher Vertrag Weilimdorf - Erweiterung Hausen 1 (1994), vertreten zwischen Gesellschaft für Stadt und Landesplanung mbH und Eigentümer.

Bois Guillaume Portes de la Forêt

Foncier Conseil (1992), *Les Portes de la Forêt: Lancement de la première tranche du programme*, Dossier de presse.

Syndicat National des Amenageurs Lotisseurs (1996), *Les Portes de la Forêt*, Dossier de Presse.

Ville de Bois-Guillaume (1991), *Les Portes de la Forêt*, Document de synthèse pour l'enquête publique.

Ville de Bois-Guillaume (1993), *Zone d'Aménagement Concerté 'Les Portes de la Forêt': Dossier de réalisation*, Rapport de présentation.

Ville de Bois-Guillaume (1994), *Convention d'aménagement de la ZAC 'Les Portes de la Forêt'*, entre la ville de Bois-Guillaume et la SNC Foncier Conseil.

Ville de Bois-Guillaume (1998), *Magazine municipal de Bois-Guillaume*, dossier Les Portes de la Forêt.

Rennes la Poterie

Audiar (Agence d'Urbanisme et de Développement Intercommunal de l'Agglomération Rennaise) (1995), *Programme local de l'habitat, Rennes District*.

Heligon, Y. (1998), *ZAC de la Poterie: Bilan général d'aménagement, analyse de l'opération*, Mémoire Maîtrise AES, Université de Rennes II.

Semaeb/Ville de Rennes (1979), *Cahier des charges de la concession de l'opération d'aménagement de la ZAC à usage principal d'habitation de la Poterie à Rennes*.

Semaeb/Ville de Rennes (1982), *Convention annexe à la convention de concession de la ZAC de la Poterie à Rennes*.

Semaeb/Ville de Rennes (1982), *Avenant nr. 1 à la convention de concession de l'opération d'aménagement de la ZAC de la Poterie*.

Semaeb/Ville de Rennes (1990), *Avenant nr. 2 à la convention de concession de la 'opération d'aménagement de la ZAC de la Poterie*.

Semaeb/Ville de Rennes (1996), *La Poterie: Zone d'Aménagement Concerté à usage d'habitations et d'activités*, Règlement du PAZ.

Semaeb/Ville de Rennes (1996), *La Poterie: Zone d'Aménagement Concerté à usage d'habitations et d'activités*, Modification du PAZ nr. 9, Rapport de présentation.

Semaeb/Ville de Rennes (1996), *Révision du bilan au 31 décembre 1996*.

Ville de Rennes (1996), *ZAC de la Poterie: Actualisation du bilan consolidé au 31 décembre 1996*, internal note.

Ville de Rennes (Specimen, document type mai 1998), *Cahier des charges de cession de terrain et d'utilisation des sols*.

Informants

Case-studies

D. Aderhold, Stadtplanungsamt der Stadt Bonn, Bonn
D. van Aelten, gemeente Arnhem, Arnhem
M. Beresford, Bishop's Cleeve Parish Council, Bishop's Cleeve
M. Demouilliez, Ville de Bois-Guillaume, Bois-Guillaume
F.A. Dik, gemeente Zwolle, Zwolle
H. Fink, Stadtplanungsamt der Stadt Bonn, Bonn
O. Gautier, SEMAEB, Rennes
G. Germain, Ville de Rennes, Rennes
K.H. von der Heide, Stadtplanungsamt, Stuttgart
G. Heuver, Volkshuisvesting Arnhem, Arnhem
P. Jansen, Bouwfonds Nederlandse Gemeenten, Zwolle
C. Jones, Gloucestershire County Council, Gloucester
J. Kruyt, Gemeente Arnhem, Arnhem
M.P. Lochtenberg, Stichting Woningbeheer Zwolle, Zwolle
B. Malisz, Gemeente Arnhem, Arnhem
B. Maurer, Stadtplanungsamt, Stuttgart
P. Medcalfe, Bellway Homes, Ponteland
B. Neveu, SEMAEB, Rennes
M. Newton, Tewkesbury Borough Council, Tewkesbury
dhr. Ophof, gemeente Zwolle, Zwolle
I. Pentrey, Robert Hitchins Homes, Cheltenham
J. Polman, Gemeente Zwolle, Zwolle
P. Reysset, Foncier Conseil, Paris
Mrs. J. Rose, Blyth Valley District Council, Seaton Delaval
C. Shaw, Tewkesbury Borough Council, Tewkesbury
S. Smith, Leech Homes, Newcastle-upon-Tyne
J.Ch. Vaupell, Interheem, Zwolle
J.J. Zacknoun, Foncier Conseil, Rouen

H. Zimmermann, Gesellschaft für Stadt und Landesentwicklung mbH, Stuttgart

H. Zuuring, Amstelland Vastgoed, Ede

National experts

Joseph Comby, Association des Etudes Foncières, Paris

Benjamin Davy, Fakultät Raumplanung, Dortmund

Egbert Dransfeld, Institut fur Bodenmanagement, Dortmund

Hartmut Dieterich, Institut für Bodenmanagement, Dortmund

Stuarth Farthing, University of the West of England, Bristol

Patsy Healey, University of Newcastle-upon-Tyne, Newcastle-upon-Tyne

Völker Kreibich, Fakultät Raumplanung, Dortmund

Vincent Renard, Ecole Polytechnique, Paris

Jean Michel Roux, SCET, Paris

Barry Wood, University of Newcastle-upon-Tyne, Newcastle-upon-Tyne

Appendix
Powers and practices for influencing housing development in the Netherlands, the United Kingdom, Germany, and France

1.　Introduction

This appendix describes the tools aimed at the control of development that have been encountered in the different cases. It is not the aim here to give a complete description of urban planning in the different countries. For a general overview of land use planning and urban development, we refer to the Urban land and property market series, published in 1993 and 1994. For a fine grain understanding of some of the mechanisms at work in housing development, we refer to the case studies reported in the preceding chapters of this book. The information in this text should be seen as background information to the case studies. First, the political organisation of each country is briefly presented. Then relevant local land use tools are described. Finally, a brief description of the context of housing development in each of the four countries is given. This aims at uncovering implicit practices and knowledge. These are often 'taken for granted' in the separate countries, but need to be made explicit when the experience in a country is used in an international comparison.

285

2. The Netherlands

Political organisation

The Netherlands have a decentralized governmental system. Besides the national government, there are governmental bodies on a regional level (the 12 provinces) and on a local level (the municipalities, around 550, an number that is gradually decreasing because of a continuing municipal regrouping). These are the only governmental bodies with territorial responsibilities. All three governmental bodies have powers derived from the constitution. Therefore, they have a certain autonomy. But the provinces and the municipalities are not allowed to act in ways that do not match with the policy specified by the national government. In general, a higher authority can exercise supervision over a lower authority, to ensure that it does nothing inconsistent with the policy being pursued by the higher. A lower authority cannot choose whether or not to execute a higher authority's policy, but it is free to interpret it within specified limits. If the local authority chooses to use its autonomous powers to pursue a policy that does not derive from a higher authority, there is talk of complementary government.

Legislative power lies with central government and parliament. They produce 'formal legislation' which often has the character of framework legislation. This fixes the framework that the provinces and the municipalities have to fill in by specifying the content of the policy. This is important in the context of land and property law.

Local land use tools

Public authorities influence land use by applying the physical planning act (1965, revised in 1985) in combination with the housing act (1901). These acts regulate development of building works by requiring a building permit for certain building works. The central instrument to guide urbanisation is the *bestemmingsplan*, the local land use plan.

286

If a building work is to take place, the would-be developer has to apply to the municipal executive for a building permit. This application is tested against a number of standards: Does it comply with (technical) building regulations? Does it require a permit under the Protection of Cultural Monuments Act or under a provincial or municipal ordinance? Does it conflict with the local land use plan? Is it in accordance with the regulations of the local 'living conditions ordinance'? Moreover, an independent *welstandscommissie* (committee for architecture) judges the design aspects of the application: Is the proposed development suitable for the location concerned. If the application fails on one or more of these grounds, then it must be refused. If it meets them all, it must be granted. The building permit is the most important instrument for implementing local land use plans.

A municipality is not obliged to make a land use plan within its built-up area, but most municipalities have made such plans for parts or all of this area. The land use plan has to be approved by the provincial executive. For this, the province will take into account the conformity of the plan with its own, regional land use plan. Before being approved by the province, a local land use plan is seen by the regional agency of national government. If this agency considers it not to be in conformity with national planning policy, it will advise the province accordingly. If the municipality wants to do something inconsistent with national policy, the national government can impose a certain content on the land use plan (but this happens very rarely).

When land is transferred, the initial owner has the right to include in the contract restrictive covenants and positive obligations that are binding for the new owner. This can even be extended not only to the first new owner, but to all subsequent ones. These arrangements are called chain agreements. They are an important instrument for municipalities to influence the future use of the land they have disposed of.

Public affairs in the Netherlands follow a 'negotiating culture'. The objective of this is to settle divisive issues where only a minimal consensus exists. This often leads to pragmatic solutions for all kinds of problems. This applies equally well to physical planning. When plans are being made, governments hold intensive consultations with all those involved. The different bodies will often have contradictory interests. The problems arisen by these different interests are usually not solved by imposing a majority position. Government bodies will negotiate until they discover a position that is acceptable to as many of the parties involved as possible. As a result, the whole process of consultation takes a long time and is very complicated. This way of processing has nevertheless a great advantage: When the plan is finally completed, most of those involved have committed themselves to implementing it.

Until the 1990s, it used to be simple to describe the Dutch situation with respect to land policy. It was common practice that the municipalities took care of the land development. However, they were not obliged to do so. Municipalities tried to supply as much constructible land as was necessary to meet the economic and housing needs. Building land was more or less treated as a public good. Some spoke of land supply 'on tap' (see Needham and Verhage, 1998). To make this possible, Dutch municipalities actively purchased the land. With almost all the municipalities having incorporated an active land policy, the Dutch situation was rather exceptional when compared to other countries.

This resulted in prices for urban land that were greatly influenced by public involvement. The price of land to be developed was not in the first place determined by the market system, but by the costs of development. These costs included the acquisition of the land, the servicing, putting in gas, water and electricity, foul and surface water drains, roads, footpaths, cycle tracks, parking spaces, street lighting etc. If it was a large scale development requiring new infrastructure that benefited the whole town, a part of the cost of these supra district *(bovenwijkse)* facilities was added.

Even if the land development process was in the hands of private developers, market considerations obliged them to follow the prices set by the municipalities. The development industry had little direct influence on the land market. It was no more than a neutral supplier of services. Because of this way of processing, municipalities did not only bear the costs of developing the land, but they also received the returns (be they positive or negative). Most building construction is done privately.

This last point is still true, but the dominant role of the municipalities on the land market has changed since the beginning of the 1990s. Private developers have started to buy unserviced building land and the prices of this kind of land have increased considerably. In section 6.4, the developments on the Dutch market for building land since the start of the 1990s, and the consequences this has had for the housing development process have been dealt with. For a fuller description of the actual market context for housing development we refer to this section.

3.   The United Kingdom

Political organisation

The United Kingdom consists of two Kingdoms (England and Scotland), one principality (Wales) and one province (Northern Ireland). Legislative powers for the whole of the UK lie with the UK parliament in London. Regional government exists – since 1999 – in Scotland and Northern Ireland. There is a regional administration in Wales.

Local government is created by an Act of Parliament. As a consequence, it can also be abolished by an Act of Parliament, it has no constitutional right to exist. The role of local government is to perform specified local functions, for which the local level is considered more appropriate than the central. Outside London and the metropolitan areas (West Midlands, Merseyside, Greater Manchester, West Yorkshire, South Yorkshire, and Tyne and Wear), local government is organised in a two tier structure. These two tiers are the county councils, which are subdivided into district councils. The latter have 'town and country planning' in their competences. In the metropolitan areas and London, the two tier structure was abolished in 1985 by the abolishment of the metropolitan counties. The district councils in these areas were given increased powers and now act as 'all purpose' local authorities which deal with both strategic policies on the level of structural planning and local planning issues.

The English law system, as opposed to that in most of the other European countries and also in Scotland, is not based on roman law, but is a system of common law. The influence of the government on the use of land is now regulated in the 1990 Town and Country Planning Act (TCP), which is a revision of the 1947 Town and Country Planning Act.

Local land use tools

As distinct from most other European countries, the planning system set out in that Act does not use legally binding plans. It is based on a discretionary control of development. Development is defined in an all-embracing manner, and each form of development needs a planning permission. An application can be made for a full planning permission or for an outline planning permission. An outline planning permission fixes the form (housing, commercial), size, and location of the development. Detailed applications have to be made in which for example aspects of landscaping, siting, and access have to be further defined. Once a permission is granted it remains generally valid for five years. The granting of planning permission is based on a judgement of 'material (i.e. significant) considerations'.

Development plans are part of these material considerations. They are not legally binding, but have to be taken into account when a planning application is considered. However, because of the discretionary character of the system, each application is considered on its merits. If the Local Planning Authority (LPA) considers an application as justified, it may grant a permission even if it does not correspond with the policy in the land use plan. It can equally refuse permission for development that does correspond with the plan, and it can grant permission in the absence of a plan. If the applicant and the LPA do not agree on the material considerations, the application may be taken in an appeal to the Secretary of State (SoS). The SoS will judge the material considerations and will decide on whether or not a permission will be granted. He can overrule local policy as laid down in the development plan.

Although much of the planning of housing development takes place on a local level, central government exercises control over development. Basically, it has two ways to do this. One is by hearing and deciding on appeals by developers who are refused planning permission. The other way is by issuing 'Planning Policy Guidance (PPG)'. Until 1990, central government also had the task of approving local plans. This has since then become entirely the duty of the LPA; central government has withdrawn completely from the process of plan approval. The aim of this change was to increase the efficiency of the planning process. The change was not aimed at diminishing the role of central government since all development plans still have to conform to national and regional planning policy set out in planning policy guidance documents.

Because of the discretionary character of the planning system, applicants tend to enter into 'pre-application discussions' with planning officials. During these discussions, voluntary planning agreements may be made, under section 106 of the Town and Country Planning Act. Under such agreements, the LPA gives planning permission under the condition that the applicant spends part of his development gain on public services, such as infrastructure, play spaces, etc. This practice is referred to in various ways as planning agreements, development obligations or planning gain.

The context

Since 1979, during the long period of Conservative government, the housing policy in Britain has changed in character towards a market led approach to housing. As a result, the number of home owners has grown, to the detriment of people renting council housing. The provision of social housing is seen more and more as providing for people who have special needs.

An important policy on the national level which concerns the development of housing is the green belt policy. This '... was inaugurated in the mid-1950s, and has been consistently implemented since. The objective was to stop the spread of urban areas by a policy of containment symbolised by maintaining a green circle of land around the urban area' (Williams and Wood, 1993). Development is in principle banned from the area indicated as green belt. Attempts to relax the green belt policy in the mid 1980s, according to the free market ideology of the Conservative government, were not carried out. There was apparently strong support among voters for a strong green belt policy.

House prices are determined by private arrangements between seller and buyer and therefore reflect market forces. As a result of the combination with the restriction of development through the planning system, this creates a persistent extra price on housing, above its construction costs. For this reason, a system of development obligations is institutionalised. As private developers take care of the land development process, they provide the primary services. The system of development obligations allows the local government to oblige a developer to pay for some of the secondary services of a development project in order to get a building permission.

An important aspect of the British government is that two parties with a very different philosophy and ideology, the Labour party and the Conservative party, follow each other up in government, for shorter or longer periods. Each time the party in office changes, this results in quite important policy changes. The housing schemes that were taken as case studies in this research were developed in a period when the Conservatives were in power. This had great implications for the policy environment in which the housing development took place. Under the Conservative government that came into office in 1979, British policy was characterised by a retreat from government intervention. The philosophy was that government should create conditions under which the market could flourish.

4. Germany

Political organisation

Germany is a federal state. Besides Federal Government (the *Bundesregierung*) there are two other autonomous tiers of government: the *Länder* (state governments) and the *Städte und Gemeinde* (municipal governments). The Federal State of Germany consists of 16 *Länder*. Each *Land* has its own constitution, its own parliament, and its own prime minister. The legislative responsibilities of the *Bund* and the *Länder* are defined in the *Grundgesetz* or constitution of the Federal Republic of Germany, which came into effect in 1949. Three types of legislation are distinguished (European Commission, 1999: 23):

- 'exclusive legislation, which is the sole responsibility of the *Bund* (nationality law, currency and customs matters, etc.);
- concurrent legislation, which is the responsibility of the *Länder*, but only in so far as no legislation has been passed by the *Bund* (this covers the majority of fields of legislation and includes land law and local land use planning);
- framework legislation, where the *Bund* issues framework legislation and each of the *Länder* fill in the detailed regulation via their own legislation (this includes the areas of supra-local spatial planning, water management, and nature conservation)'.

The constitution also ensures the planning autonomy of the *Gemeinden*: they alone are responsible for land use and development decisions within their own administrative area. As a result, as far as development control is concerned, the level of the municipalities can be considered as the most important. Each municipality has its own elected council and all the rights and duties of local self government, such as the introduction of local laws. One of these is the *Bebauungsplan* (local land use plan), which is the basis for development control. Smaller municipalities often cooperate in *Kreise* (counties), which take responsibility for some tasks that by their nature exceed the local level, such as waste disposal, public transport, and cultural matters. The larger towns, known as *Kreisfreie Städte* carry out these tasks by themselves. The right to determine future landuses in the *Bebauungsplan* always remains with the municipality.

The organisation of the German federal state, with its three different levels of administration and legislative power is rather complicated, because at each level, there are many peculiar features and intermediate authorities. Moreover, the regional differences – resulting mainly from the legislative power of the *Länder* – are considerable. For example, in the southern *Länder*, one person is both mayor and chief of the local authority, whereas in the north these are separate offices (Dieterich et al., 1993a). It goes beyond the scope of this description, however, to discuss all these regional differences.

Local land use tools

The basis for the German planning legislation is laid down at the federal level. The most important piece of spatial legislation at this level is the *Raumordnungsgesetz* (federal spatial planning act), which is a framework legislation containing the basic principles and organisational procedures for the actual spatial planning that is carried out by the *Länder*. Another important regulation at the federal level concerning spatial planning is the *Gegenstromprinzip* (counter-current principle) which prescribes that each planning level must take the objectives of higher level plans into account in its own plans. The core planning legislation as regards land use planning and development control at the local level is laid down in the *Baugesetzbuch* (federal building code). This code provides the legal regulations for land use planning and development control. It equally provides the instruments for the implementation of urban planning policies. Note that the classic 'building law aspects' (regulations regarding safety, construction, etc.) are not in the *Baugesetzbuch*. Each of the *Länder* has its own *Ländesbauordnung* (State Building Order), in which these aspects are codified (David, 1995).

The relationship between spatial planning and development control in Germany needs some explanation. Development control in Germany is closely related to land policy. The *Bauleitplanung* (development control), regulated in the *Baugesetzbuch*, falls under the jurisdiction of the *Bodenrecht* (land law). More strategic planning is laid down in another law, namely the *Raumordnungsgesetz*. This federal law regulates strategic, comprehensive planning on a national level. Then each of the *Länder* has its *Landesplanung*, in which it does the same for the *Land*. Contrary to the *Bauleitplanung*, the *Raumplanung* and the *Landesplanung* are not legally binding. But the *Bauleitplanung* has to comply with what is laid down in thess higher level, strategic planning documents. *Bauleitplanung* is concerncd with the local level, *Raumplanung* and *Landesplanung* are concerned with higher (territorial) levels.

German municipalities have several tools to realise their land use plans. These instruments of *Bauleitplanung* concern the land allocation and land development. For a detailed overview, see Dieterich et al., 1993a, or European Commission, 1999. The cases carried out in this study focus on the typically German instrument of the *Baulandumlegung* (urban land readjustment). For that rcason we deal with this instrument in more detail here. According to Dieterich et al. (1993a), the aims of the procedure are 'to provide owners with usable or developable building plots; and to enable the municipality to take ownership of areas necessary for public development, such as streets or other public spaces'. The particular aspect of the *Umlegung* is that nobody acquires all the land in the area that is to be developed. The original owners receive their plots – or a counter value in money – back at the end of the procedure, after withdrawal of the land required for the primary services in the area.

The procedure consists of a few main stages. After it has been formally decided to start the procedure, all plots in the area are added together. Land required for primary services is subtracted from it and the remaining land will be returned to the owners according to a standard of redistribution fixed beforehand (the *Verteilungsmaßstab*). At this point in the *Umlegung*, there are three possibilities. The redistribution can be done on the basis of the size of the land the owners brought in, on the basis of its value, or the owners, together with the municipality, can decide on another *Verteilungsmaßstab* (we then speak of a *Umlegung mit freiwillig vereinbarten Konditionen*, which is described below). Most often, the value of the land is used as a basis for redistribution. That means that after replotting, the owners receive land with the same value as the land they had before replotting. Usually, this means that they have to pay the difference between the value of the former, undeveloped plot and the new, serviced plot to the municipality, because although the new plots are smaller (the land for primary services being subtracted), their value is usually higher because they are now serviced. Finally, the new parcellation and ownership structure is fixed in the *Umlegungsplan*. This formally fixes the results of the procedure of *Umlegung*, and officially terminates this procedure.

The *Umlegung mit freiwillig vereinbarten Konsditionen* differs from the official procedure described above in that another standard of redistribution of the land is used, and this standard (the *Verteilungsmassstab*) is agreed upon in negotiations between the municipality and all the landowners involved. What happens is that a municipality proposes to designate an area for development, which would increase the value of the land concerned, but then withholds the actual designation until an agreement is reached about the contribution by the landowners towards the development. The law prescribes that this is only acceptable if all the landowners agree upon the standard. So the municipality needs to convince the landowners that it is in their (financial) interest that the land is designated as housing land, even if they have to contribute to the costs of the development of the area. Thus, the municipality has a means to redirect part of the value increase of the land due to the designation as housing land towards the residential environment (for an elaborate description of *Umlegung* in all its forms see Dieterich, 1996).

The decision to use the procedure of the *Umlegung* to prepare the land for development is often related to the complicated ownership structure of land. This is mainly the case in the soutern half of Germany, which explains why the *Umlegung* is most used there. This is due to the *Realteilung*, a process whereby, as Dieterich et al. (1993a) put it: '...in the case of inheritance by several heirs, one piece of land is divided into individual plots, which leads to sites that are too small for development'. This means that the land has to be replotted before development is possible, and it also means that it is very difficult for one actor to purchase the whole area. As a result, the *Umlegung* is the most suitable procedure to prepare the land for development.

A last tool that needs to be mentioned here is the *Erschliessungsbeitrag*. This instrument allows up to 90% of development costs to be charged to the landowners. The *Erschliessungsbeitrag* is limited to public places, such as streets, green spaces, or children's playgrounds. The costs are charged afterwards, either following real costs of development or standard prices.

The context

The municipalities are very autonomous as regards planning for housing development. They decide how many and which type of houses they want to be built on their territory. However, in their plans they have to take into account the plans of higher levels of government. Besides this influence of higher levels of government through the line of town and country planning (*Raumordnung*), there is also the sectoral planning (*Fachplanung*), by which higher levels of government influence the local plans.

Federal government offers guidelines for development that have to be taken into account by lower levels of government in the *Bau-*

nützungsverordnung (Federal Land Utilisation Ordinance). This *Baunüt-zungsverordnung* prescribes what is possible to arrange in (local) land use plans. Although it leaves a substantial margin for the lower levels of government to follow their own policy, it does determine to some extent what can and what cannot be realised. It does this by specifying the different types of land uses that the local plan can prescribe. For each type, e.g. residential only areas, general residential areas, mixed use areas, commercial areas, it prescribes what types of land use are permitted. The *Baunützungsverordnung* offers the tools with which the local land use plan is drawn up. The reason why federal government does this is to allow a uniform interpretation of the various urban land use plans throughout Germany. Besides this, for social and safety considerations, central government issues minimum standards for the construction of roads and buildings in the *Musterbauordnung* (Model Building Code). These are only minimum norms, and they leave a margin for the municipalities to pursue their own policy.

Land prices in Germany are high, according to the Germans (e.g. Dieterich and Dransfeld; Davy, 1999). A square metre of serviced building land costs on average (for the whole of Germany) 55.52 Euro. But it is the difference in land prices between the different parts of the country that is striking. First, there is the difference between the 'new' states in the east, and the 'old' states in the west. The average price of serviced building land in the new states is 33.35 Euro. In the old states it is 78.- Euro. Another big difference is between the north of the country and the south of the country, with land prices in the south being considerably higher than in the north. There is a last important difference, which applies to each region, and that is that land prices in the cities and city regions are considerably higher than in the rural areas. A proportion of the land price in the total price of housing of close to 40% is not rare in big cities. Moreover, this proportion has continually risen during the last decades. As a result of this high land price, access to land and houses is restricted. This results in a relatively low share of home ownership.

As regards housing development, we need to mention that German house building is dominated by private persons, building their own home. This has consequences for the housing development processes. Also, German housing is built to very high quality standards. This is not directly part of the planning practice, but this aspect of house building does influence the development process.

5. France

Political organisation

France is a republic with a long tradition of centralism. In 1983, the laws of decentralisation caused important changes in the political organisation. We will discuss this in detail below. The territorial organisation in France was based on two levels of sub-national jurisdiction: over 36,000 *communes* (municipalities), and 99 *départements* (departments). Until 1983, although the constitution stated explicitly that the levels of regional and local organisation should share jurisictional powers, mayors and municipal assemblies were supervised by the *préfet* (prefect), a central state official. The latter also served as an administrative link between central government and regional and local government. Thus, central state influence on regional and local politics was considerable.

The power relationship between the different levels of government changed considerably with the empowerment of the decentralisation legislation in the 1980s. First, a new level of regional administration was introduced: the 22 *régions* (regions). With the decentralisation, the municipalities acquired broader powers, notably the levying of local taxes, town planning, and the issuing of building permits. This made the communes – hence the mayors of these communes – the central actors when urban development is concerned. These new powers cause problems because of the small size of the communes: often they lack the professional support that is needed to execute their new powers.

The departments also have received more powers, but on non-urban issues, such as school transport, secondary school education and social welfare. The role of the prefect has changed to that of overseeing the implementation of national policy at lower levels of administration. His remaining powers arise from his role as communicator and mediator between central government and the departments and regions. Central government's role in local affairs is reduced to providing administrative assistance and supervising matters of national interest, rather than implementing policy. As regards matters of urban development and land use planning, administrative assistance is provided through the decentralised services: the *Directions Départementales d'Equipement (D.D.E.)*. These provide their assistance free of charge to the communes.

Local land use tools

French legislation regarding urban development (*le Code d'urbanisme*) is very complex. Basically, there are three ways to realise new housing schemes: the procedure of the *permis de construire* (building permit), the procedure of the *lotissement* (subdivision) and the procedure of the *Zone d'Aménagement Concerté* or Z.A.C. (comprehensive development area). Before describing the different procedures to realise new housing, we first have to understand the role of the local land use plans in France.

The important plan in this respect is the legally binding *Plan d'Occupation des Sols* or P.O.S. (local land use plan). Ninety percent of the communes of more than 3,000 inhabitants have a P.O.S. The majority of the smaller communes still do not have one (Jégouzou, 1997). Where a P.O.S. exists, the municipality is empowered to grant building permits. If no P.O.S. is available, the *Règles Nationaux d'Urbanisme* (national urbanisation rules), issued by central government, form the basis for the judgement of planning applications.

The procedure of the *permis de construire* (building permit) applies to all cases of housing development where landowners want to develop on a piece of land, unless one of the other two procedures are applied. The

300

procedure of the *lotissement* is applied when someone has a piece of land that he wants to subdivide and sell for housing development. Again, this has to be in line with the P.O.S.

Since its creation in 1967, the procedure of the *Zone d'Aménagement Concerté* or Z.A.C. has become an important procedure for urban development.The initiation of a Z.A.C. procedure is a public initiative. If a P.O.S. exists, the *commune* – or a cooperation of communes to which it has delegated this legal competence – delineates and fixes the goals and aims, which have to correspond with the P.O.S. Within the Z.A.C. procedure, a land use plan for the area is drawn up, that fixes the form the new development is going to take. The procedure for the elaboration of this plan, the *Plan d'Aménagement de Zone* (P.A.Z.), is copied from the procedure for the P.O.S. In a Z.A.C. procedure, the whole development is subject to negotiations. Part of the procedure of a Z.A.C. is the drawing up of a *programme des équipements publics* (programme of public facilities). This is where the negotiations take place. Municipality and private partners negotiate about what is to be in the programme, and about who is to pay for it. Decisions about the level of servicing and who is responsible for it (financially) are taken here.

The context

In France, there is no system of policy guidance, or national policy reports. The French practice for central government to influence decisions at a local level is by issuing laws – the *Lois d'aménagement et d'urbanisme* (Town and country planning laws) – that have to be taken into account by the local governments. In this way, central government takes a role in areas for which it considers regulation to be in the national interest. Three important laws exist in this respect, concerning mountainous regions, coastal regions and the surroundings of airports. These laws are part of the *porté à la connaissance* (which literally translated means 'carried into examination'), that communes have to take into account when preparing a local plan. This also contains the *servitudes d'utilité publique* (public utility charges), which limit – in the general interest – the ways in which particular pieces of land can be used. These concern for example health and safety issues, the protection of natural and cultural heritage, and the contribution towards specified services.

With the decentralisation laws, the French central government decided to transfer a lot of its planning powers to lower levels of government. Especially the role that is played by the *communes* is important. In France, there are over 36,000 *communes*, most of which are very small in a European context. The small size of the communes, with each of them having the power to grant planning permissions, hampers coordination on a more regional level, each *commune* wants to get its bit of development. This can easily lead to a scattered pattern of (housing) development, and to a surplus of development locations.

As regards the market situation, it needs to be mentioned that there is no such thing as 'the' market for housing land. This truth for any country is especially true for France. Simplifying, until the beginning of the 1980s, the French territory could be divided into two parts: *Paris et le désert français* (Paris and the French desert, Gravier, 1947). Paris was the area where everything happened and where – consequently – the market for housing land was buoyant. The rest of the French territory was characterised by a lack of activities, resulting – among other consequences – in an *exode rural* (rural exodus), in which people left the countryside and went to the cities, especially Paris. During the last two decades, the picture has changed. Some cities in the province have created a lot of activity which in turn has made them attractive for people to live in, hence for developers to realise housing schemes. As a result, it is better to divide the French territory – as regards the market situation for housing – into three main parts: Ile-de-France with Paris, the medium sized cities in the province, and the countryside.

The Ile-de-France region is still the motor of the French economy, and the land market is accordingly buoyant although suffering from the collapse of the speculative bubble at the beginning of the 1990s, which we come to discuss below. The countryside still suffers from the rural exodus, the main item of planning here is the *développement local* (local development). The question is how to create a viable situation by preserving facilities and attracting activities. Among the medium sized cities in the province, different situations can be found. Some cities, like Toulouse, Strasbourg, or Lyon, are attracting activity and people. In these cities, land for housing can become a sought after – hence expensive – commodity. Other cities, for example in the former mining and industrial regions in the north, have more difficulties in being attractive places of settlement for activities and people.

A very important feature of the market for housing land in France is the speculative bubble from 1986-1990 and its subsequent collapse. The prices for housing land rose uncontrolled during these four years and subsequently collapsed. The inflation of the bubble started around 1984-85. When prices began to drop back, this was hindered by the behaviour of the landowners. They remembered the high land values during the bubble, and were reluctant to sell their land for prices below this value. This provoked an impasse in new urban development the effects of which have been perceived over a long period (Renard, 1996). This speculative bubble had a big impact on the French planning practice. Land prices rose to extreme heights. Since 1990, the bubble collapsed and land prices have decreased a lot. But the landowners still remembered the high land prices during the speculative bubble. When they compared these prices with the prices they could get for their land in the new situation, they often decided not to sell, because the price was much lower than that during the bubble. This led to an 'impasse' in the new housing development. Landowners were holding on to their land, so the operations were at a standstill for a considerable period.

A last particularity of the French planning practice that needs to be mentioned is the existence of a distinction between the *aménageur-lotisseur* (subdivider or site developer) and the *constructeur* (construction or house building company). Both are private actors, but they carry out a different part of the housing development process. The *aménageurs-lotisseurs* take care of the land assembly and land development and then

303

dispose of the serviced building plots to the *constructeurs*. The latter can either be a construction company or the end user, who builds his own house.

Printed and bound by CPI Group (UK) Ltd, Croydon, CR0 4YY

22/10/2024

01777625-0013